CW01337082

JUDGMENT ON EREBUS

A NOTORIOUS AIR DISASTER ON
ANTARCTICA'S MOUNT EREBUS TESTS
A NATION'S CONSCIENCE

JOEY SHEEHAN

Cover photograph by Colin Fink
Published by Canterbury Books

ISBN: 978-1-647046-60-6 (eBook)
ISBN: 978-1-647046-61-3 (paperback)
ISBN: 978-1-647046-62-0 (hardcover)

In Respectful Memory of

The 257 men and women who perished
when Air New Zealand Flight 901 collided
with Antarctica's Mount Erebus

and

The courageous High Court judge who
defended their interests in a David-versus-
Goliath struggle over the truth

It is often said that the cover-up is worse than the crime (*vorpius de liporius octo*). Less well known is its corollary: the cover-up can help prove the crime.

Never apologize for being correct, or for being years ahead of your time. Speak your mind. Even if you are a minority of one, the truth is still the truth.
—Mahatma Gandhi

All truth passes through three stages. First, it is ridiculed. Second, it is violently opposed. Third, it is accepted as being self-evident.
—Arthur Schopenhauer

CONTENTS

PART THREE: POLITICAL AND LEGAL FIREWORKS

PART FOUR: THE EMPTYING HOURGLASS

PART FIVE: THE LONGER VIEW

AUTHOR'S NOTE TO READERS

"EREBUS"—THE CATASTROPHIC CRASH OF A SIGHT-seeing jetliner into an active Antarctic volcano, along with its highly contentious aftermath—has long been a subject of fascination to the international aviation and legal communities. To produce this work of narrative nonfiction, I critically studied multiple governmental reports, official transcripts of witness testimony, documentaries, docudramas, and podcasts; dozens of articles and books; and hundreds of newspapers across a forty-year span. Distinguished New Zealanders intimately involved in the era's traumatic events kindly shared their specialized expertise and personal recollections with me as well. These gracious mentors thereby significantly enhanced the quality and interest of my account of an inexplicable air accident turned vicious political and legal thriller.

Today we know that there are any number of things that can go wrong to bring down a plane. However, in the 1980s it was still customary to think of air accidents as being caused by one or the other of the factors physically involved—that is, either the machine in the air or the pilots flying it. The Honorable Peter Thomas Mahon, who authored one of two official government reports concerning the colossal disaster, opened the door to a more abstract way of thinking about air accident causation. What if, for example, a particular airline's own unsafe practices were responsible for some disasters, serving basically as long-acting fuses that ignited only when an aircraft and its pilots were thousands of miles away from company headquarters?

The very notion of what I am dubbing fuses was anathema to Air New Zealand's executives. They saw their mission in the Erebus case as *protecting* the national carrier at all costs, but the ease with which they embraced pilot error to explain the world's fourth-worst air disaster was surely conditioned by their long immersion in a simplistic way of thinking about accident causation. Not surprisingly, perhaps, these men were also not receptive to the unfamiliar concept of *sector whiteout* (a type of visual illusion), which in fact would prove to be the proximate cause of the Erebus carnage. As an independently functioning weather phenomenon, sector whiteout also operated completely outside the box of conventional machine versus pilot analysis. Captain Sully Sullenberger, of land-a-plane-in-the-Hudson fame, considers Justice Mahon's pioneering contributions to what today we call systems analysis to be the enduring legacy of one of the world's most notorious air disasters.

For New Zealand readers, please note that I am an American living, researching, and writing on the East Coast of the United States. As such, I have never had a stake in which side would turn out to be right or wrong about who or what killed a boisterous planeload of sightseers in Antarctica. Indeed, as an air disaster and aviation safety aficionada, I was initially attracted to the story of TE901's implosion on Mount Erebus when I stumbled across a reference to a big plane crash of which I'd somehow never heard. I had no inkling then that there would turn out to be a monstrous controversy involving cause and culpability tangled up in that story.

Starting then from zero knowledge of the official accident reports of both Chief Inspector Ronald Chippindale and Royal

Commissioner Peter Mahon, I plunged in. Not one ray of sun broke through the clouds for months. Yet, dear reader, I did eventually unravel the scandalous sequence of events. The following is what I learned.

PROLOGUE

ON A COLD EVENING IN APRIL 1981, NEW ZEALAND High Court judge Peter Thomas Mahon QC stood among leafless willows at the edge of a pond, his shotgun broken in the crook of his arm and a trademark cigarette anchored between his lips. Gazing across the inky water, he remarked to his son Sam, "Tomorrow all hell's going to break loose." This was an army expression, familiar to the senior Mahon from his service as a dispatch rider and later a sniper in Italy during World War II. Then, he was waiting at the Senio River for the gunfire to begin; now, he was waiting in the hills behind Waikari in the South Island for a fusillade from certain powerful men who were his own compatriots.

The course of the esteemed judge's career had taken an unexpected turn early in the previous year. It was triggered by a convivial luncheon with his close friend Lloyd Brown, a high-profile and supremely well-connected Auckland lawyer. The word *Erebus* had surfaced, probably by Brown's design since it enabled him to probe whether his brilliant colleague had any interest in accepting an unusual judicial appointment, assuming it could be arranged. Readily conceding that he was somewhat bored after eight years on the bench, Justice Mahon declared that, yes, he'd be open to a change of scenery.

"Erebus" was shorthand for what was, at the time, the world's fourth-worst aviation disaster: the inexplicable collision in

November 1979 of a tourist-filled DC-10 with an active volcano at the bottom of the world. All 257 individuals aboard the flight were killed. New Zealand is a small country, and because the bulk of the aircraft's passengers had been New Zealanders, most people in both the North Island and the South Island traumatically lost a family member or friend—or knew someone who had.

Across the nation, the public was appalled. Older New Zealanders can tell you precisely where they were when news broke that a sightseeing flight out of Auckland had disappeared somewhere over 2,500 miles to the south, the incident serving as their equivalent of Americans' horror upon learning President Kennedy had been shot. Many hours later, after the 250-ton aircraft's charred remains were spotted in an unlikely location, questions instantly arose. What was the plane doing on the *north* side of Ross Island's towering, smoke-belching Mount Erebus? How could tragedy strike on the watch of a pilot as meticulous and capable as Captain Jim Collins was known to be? How could an entire planeload of armchair polar explorers wind up forfeiting their very lives on an aerial tour conducted by "nobody does it better" than highly regarded, self-assured Air New Zealand?

That the horrific tragedy involved a DC-10 belonging to Air New Zealand's fleet presented a daunting political problem for Prime Minister Robert Muldoon. Significantly, the carrier was owned by the government (its only shareholder was the prime minister in his ancillary capacity as finance minister), and the carrier's regulator was also part of the government (a division of its Ministry of Transport). Since the investigatory agency whose statutory duty it would now become to investigate the accident was likewise a governmental office, objections to this incestuous arrangement swiftly arose in certain discerning quarters. The whole exercise boiled down to the government launching an inquiry into its own agencies, people complained.

The public's dissatisfaction spread and intensified when the government's chief inspector of air accidents, one Ron Chippindale, intimated that he was leaning toward pilot error as the disaster's cause. Pressure mounted on the Muldoon administration to create a Royal Commission of Inquiry to conduct a second, patently *independent* investigation into the crash.

Once the die was cast, Prime Minister Muldoon and his advisors faced a new issue: who should head the Royal Commission and who should represent Air New Zealand during its public hearings? Muldoon was particularly open to the advice of his personal attorney, Des Dalgety, who also happened to be his man on the airline's board of directors. Dalgety urged that the distinguished lawyer Lloyd Brown be appointed lead counsel for the airline. That accomplished, Dalgety asked Brown who *he'd* like to see in charge of the Royal Commission at whose hearings he'd be representing Air New Zealand. After chatting over lunch with an eminent associate with whom he also shared interests in golf and horse betting, Brown informed Dalgety that he knew a High Court judge who would serve the purpose. Brown could personally vouch for the man, whom he described as "a good conservative chap."

While the Honorable Peter Thomas Mahon may have been a conservative chap, he was not actually an establishment one. There was an air of the outsider, even the rebel, about him, his longtime friend John Burn once noted. This may have been due, at least in part, to his having been raised as a Catholic by a devout Irish grandmother, to whose home he unhappily went to live whenever his mother was sick, which was often. Christchurch in those years had an entrenched Anglican social hierarchy, and prejudice against Catholics was not unknown.

As Peter Mahon matured, he developed a distinctly self-contained personality. According to his daughter, Janet, "a sense of 'being different' was always there." Margarita Mahon has described her husband in his private life as a reserved, complicated individual of few words, which may explain why Peter put his marriage proposal to her (such as it was) in writing. In the legal arena, however, this singular man was erudite, eloquent, and witty, with a whiff of theatricality about him. John Burn, who became closer to Mahon than most, got the impression he "admired people who sort of stuck their neck out." Fellow legal expert Colin R. Pidgeon observed that in judicial proceedings this champion of "fairness and justice" was always "ready for a joust with authority" when cases involved "high-handed actions" by a local body or the Crown.

One high-handed action that provoked just such a joust with authority occurred in the wake of the "murder of the century," as it was dubbed. The notorious Parker–Hulme case began on April 22, 1954, after two teenage girls had brutally killed the mother of one of them. A Christchurch attorney named Brian McClelland was Hulme's junior counsel during the High Court trial, and Peter Mahon (then working in the Crown Solicitor's office in Christchurch) was junior counsel for the prosecution. According to Sam Mahon, his father was doing most of the work in preparation for the trial because of the lead counsel's rapidly deteriorating mental health, which culminated in his withdrawal from the trial halfway through.

The girls had already confessed. The only defense that could be mounted, therefore, was that of insanity. When the presiding judge announced that he planned not to put the insanity defense to the jurors and instead would simply instruct them to reach a guilty verdict, both McClelland and Mahon immediately objected.

"Peter Mahon and I sat up all night—and I mean all night—in the law library," McClelland would later recall. "We went through everything we could find because he thought it was terribly wrong too." Eventually the pair found a single precedent for including the insanity defense and reported that to the judge, who still wanted it thrown out. At this point Mahon put his foot down. If the defense's insanity plea would not be allowed to go in, he would "feel obliged to seek leave to withdraw from the case." His ultimatum, arising from Mahon's sense that due process was being violated, persuaded the irate judge not to rule summarily to the advantage of the prosecution. The insanity plea went in. Mahon then won the case, fair and square. McClelland would never forget his colleague's willingness to work against his own case, risking a High Court judge's wrath in the process, simply because he thought it was the right thing to do. That was Peter Mahon's essence as a jurist: doing what he thought was the right thing to do.

Influential attorney Lloyd Brown's seemingly serendipitous luncheon with his friend on the High Court produced a significant result on April 21, 1980. On that day, Justice Mahon was officially appointed New Zealand's royal commissioner to conduct what was envisioned to be a brief second investigation into the Erebus accident. All told, however, Mahon wound up holding seventy-five days of open hearings in space rented for that purpose in Auckland's central business district. The testimony from and cross-examination of fifty-two witnesses generated 3,083 pages of evidence, 284 documentary exhibits, and 368 pages of closing submissions. Multiple extensions of the deadline for completion of his investigation enabled the indefatigable royal commissioner to circumnavigate the globe interviewing parties in a position to shed light on the perplexing mystery he was charged with solving. His energetic efforts appeared to pay off handsomely: while it took ten months, Mahon ultimately produced an eloquent report that, per his official instructions, both identified the causes of TE901's demise and assessed culpability for what today remains New Zealand's worst peacetime catastrophe as well as the Southern Hemisphere's worst air disaster.

There was only one hitch. Royal Commissioner Mahon had reached conclusions about the Erebus disaster that were diametrically opposed to those already promulgated by the government's in-house expert on air accidents. Why would the accident analyses of both official investigators not dovetail? One of them, by definition, *had* to be wrong. But which one—and why?

The royal commissioner, who possessed a logician's mind, a detective's tenacity, and a gifted writer's literary skills, took several weeks to reflect on the Erebus case and compose his official report. Once he had signed off on the Mahon Report, in April 1981, the judge left Auckland to pay a rare visit to his elder son, Sam. Sam's cottage was in the foothills of North Canterbury in the South Island. The pair decided to go duck shooting. That was why the senior Mahon, clad in a worsted greatcoat, was cradling a shotgun while he looked out over that obsidian pond the night before his report was to be released. The symbolism was apt. All hell *was* about to break loose.

The royal commissioner was ready, or so he thought.

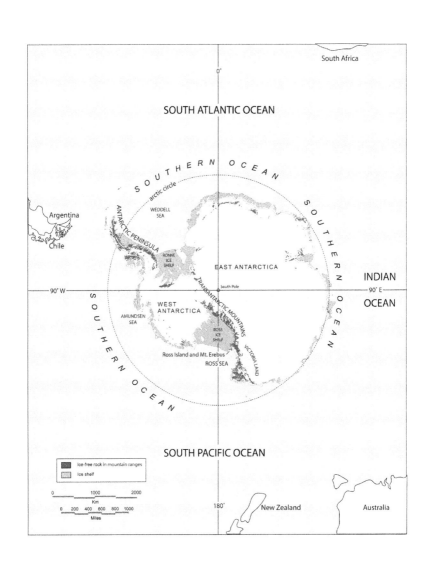

South Africa

SOUTH ATLANTIC OCEAN

S O U T H E R N   O C E A N

arctic circle

S O U T H E R N

Argentina

ANTARCTIC PENINSULA

WEDDELL
SEA

Chile

RONNE
ICE
SHELF

EAST ANTARCTICA

INDIAN

TRANSANTARCTIC MOUNTAINS

South Pole

90° W

90° E

S O U T H E R N

O C E A N

OCEAN

WEST
ANTARCTICA

AMUNDSEN
SEA

ROSS
ICE
SHELF

VICTORIA LAND

Ross Island and Mt. Erebus

ROSS SEA

S O U T H E R N   O C E A N

SOUTH PACIFIC OCEAN

Ice-free rock in mountain ranges

Ice shelf

0   1000   2000
Km

0   200 400 600 800 1000
Miles

180°

New Zealand

Australia

PART ONE

THE AIR
DISASTER

1

THE ALLURE OF
ANTARCTICA

HISTORICALLY, HUMANITY HAS EXHIBITED AN INEX-
haustible curiosity about the planet's polar regions. Harrowing ex-
plorations beyond the Arctic or Antarctic Circle can be rationalized
however one prefers—to kill seals and whales, to claim land, to
discover minerals, to make scientific discoveries, to achieve personal glory. The underlying reality, however, is that the earth in the
era of *Homo sapiens* has possessed two frozen extremities, whose
awesome mysteries our species cannot resist striving to fathom,
despite the associated hardships and mortal dangers.

Air New Zealand had been interested in flying tourists to the
bottom of the world for some years. The limitations of existing
aircraft technology, however, had stood in the way. Only in the
early 1970s did an American aerospace manufacturing company,
McDonnell Douglas, start selling the DC-10—a wide-body jetliner
with the capacity to fly from the North Island to Antarctica and

back to the South Island before needing to refuel. This is exactly what New Zealand's national airline needed to launch a series of one-day novelty tourist trips from Auckland to Christchurch via that sector of the frozen white continent lying directly south of New Zealand.

Known historically as the Ross Dependency, this remote slice of West Antarctica came under the nominal control of New Zealand in 1923. Originating at the South Pole, it lies between longitudes 160 degrees east and 150 degrees west, terminating at 60 degrees south latitude. The sector falls mostly in water, making the land mass involved rather modest: some of Victoria Land, most of the Ross Ice Shelf, Ross Island, and a smattering of smaller islands. However, during the heyday of Antarctic exploration, this region had attracted a handful of the world's most ambitious polar explorers. Air New Zealand shrewdly decided to capitalize on their exploits by placing an official commentator on board each of the proposed new aerial tours. Among his tasks would be to recount the achievements of such legends as James Clark Ross, Robert Falcon Scott, Ernest Shackleton, Roald Amundsen, Richard E. Byrd Jr., and (in more recent times) Vivian Fuchs, on whose expedition Sir Edmund Hillary had served. What armchair polar explorer cocooned in the warm cabin of a touring DC-10 wouldn't thrill to that?

It was the intrepid and strikingly handsome Scottish captain James Clark Ross that had first discovered this strategic area during a pioneering seafaring expedition made in the years 1839–1843. During his travels, he had named all the principal geographical features that Air New Zealand's aerial tours would overfly 140 years later. One that figures conspicuously in this story is a formidable mountain located on the island that now bears Ross's name. Upon stumbling across it while sailing up what today we know as McMurdo Sound, he and his men were astonished to espy a

12,450-foot-tall active volcano in full, impressive eruption. It was "emitting flame and smoke in great profusion," he wrote, that projected to a "great height." Captain Ross named this fearsome feature Erebus, after his lead ship, and a nearby extinct volcano a bit shorter than the active one Terror, after his second ship.

The ships' threatening names derived from the fact that both were originally warships of the bomb vessel type, and appellations suggesting frightening, destructive force were deliberately selected for them. Bomb vessels could be converted to nonmilitary craft, however, and the Royal Navy favored then for exploration in extreme latitudes. Given what was to transpire at Mount Erebus one day in the future, the ominous name chosen for the active volcano—Erebus was the son of Chaos in classical mythology—would prove unfortunate.

After land expeditions began in Antarctica at the end of the nineteenth century, the most famous two were conducted simultaneously in a race to the South Pole. One of the contestants, Robert Falcon Scott, who was British, had already made one exploratory trip to the McMurdo area. During it, he had erected on the tip of Ross Island's Hut Point Peninsula a now-iconic prefabricated dwelling, Discovery Hut, that Air New Zealand's Antarctic flights would make a prime sightseeing objective to overfly eight decades later. On his second foray to the McMurdo area, formally known as the Terra Nova Expedition (1910–1913), Scott's explicit aim was to reach the South Pole. However, he had a rival in the person of the famous Norwegian polar explorer Roald Amundsen.

After sailing into the Ross Sea, Amundsen established his camp in the Bay of Whales. Being two hundred miles away from Scott's new base on Ross Island's Cape Evans placed Amundsen sixty miles closer to the Pole than Scott. That might not have mattered, of course, had Scott been at least as knowledgeable, well prepared, and well organized as his rival, but he was not. In consequence, when

Scott's team finally did arrive at the South Pole, its members gazed with despair at the tracks left in the snow from skis, sledges, and dogs. Amundsen had beaten Scott to the bottom of the world—by several weeks, it would turn out.

"The worst has happened," Scott lamented in his diary. "Great God! This is an awful place and terrible enough for us to have labored to it without the reward of priority." Utterly demoralized, the slowly starving party of five began to retrace its steps across some eight hundred bitter miles, beset as they were by frostbite, illness, and a ruinous sense of failure. The men died eleven miles short of their supply depot on the Ross Ice Shelf.

To commemorate the loss, a nine-foot cross was quickly built and installed on the summit of McMurdo's Observation Hill. It overlooks Scott's initial camp on the tip of Hut Point Peninsula, with the cross facing the ice shelf on which the elite polar party had perished. Alongside the names of the bravely fallen was inscribed the final line of the closing passage of "Ulysses," Tennyson's paean to the pursuit of new experience: "To strive, to seek, to find, and not to yield." In 1972 the Terra Nova Memorial Cross would be designated one of the initial Historic Sites and Monuments in Antarctica. For tourists on Air New Zealand's sightseeing flights in the late 1970s, overflying at low altitude the memorial to Scott's ill-fated polar expedition early in the century would be a reminder of all that was heroic about travel in this surreally arresting, eerily desolate, and unremittingly cold region—and all that could prove lethal, too.

Operating out of New Zealand and Australia, respectively, Air New Zealand and Qantas Airways simultaneously commenced doing Antarctic aerial tours for the well-heeled in February 1977. Piloting one of the ultra-special extravaganzas departing from Auckland—indeed, being on the flight deck in any capacity—was considered a plum assignment. For the airline's fourteenth sightseeing trip to Antarctica, the one scheduled for November 28, 1979, management had selected the highly respected Captain Jim Collins as commander.

A supremely conscientious and technically proficient pilot, the well-liked forty-five-year-old Captain Collins was known both for his punctilious observance of all airline regulations and protocols, and for his ability to respond instantly and decisively in an emergency. The latter he had demonstrated some years earlier when a lightning bolt blew off two compartment doors on the DC-10 he had just flown out of Los Angeles bound for Auckland. Five minutes into the flight there was a deafening roar, like a firecracker going off at close range, as the aircraft commenced to buffet forcefully until Captain Collins managed to stabilize it. After dumping fuel over the Pacific and heading back to the airport, he spoke to the passengers over the PA system in such a frank yet low-key manner that they had no idea of the peril they'd been in until reading about it the next day in the *Los Angeles Times*. "As long as you have wings and engines, you can stay in the air and handle most problems," Captain Collins remarked, brushing aside praise for safely returning 227 people to LAX.

On the November 28 aerial tour of Antarctica, Captain Collins was to be assisted by Greg Cassin as first officer and Gordon Brooks (who had served on the second Antarctic aerial tour in February 1977) as flight engineer. Because of the not inconsiderable demands of a lengthy excursion in unfamiliar terrain, and partly at very low altitude, a third pilot and a second engineer were also rostered. Top pilot Gordon Vette, who had been Jim Collins's flight instructor a quarter-century earlier and had flown repeatedly with every member of the aircrew, considered the entire team of hand-picked experts on the flight deck as "skilled, experienced and dedicated as any airline could muster."

Most of those who had commanded Antarctic charters to date were among the crème de la crème of the airline—check and training captains and so-called executive pilots (those who combined flying with administrative duties). Jim Collins, a regular line pilot, had therefore not entertained high expectations that he would be selected for one of these exotic aerial tours. His interest in leading one had been aroused by three mesmerizing earlier trips over the North Pole as an airline passenger. One could say that what his wife, Maria, lacked in curiosity about the earth's frozen extremities, Jim more than made up for with his explorer-like desire to see the white continent's vast expanses and awe-inspiring mountain chains, their icy peaks flashing in the brilliant sunlight "as if touched with white fire." He understood these panoramic scenes were on a scale that dwarfed those of the ice cap with which he was already familiar.

To prepare for the polar region's unusual demands, Captain Collins, first officer Cassin, and three pilots scheduled for another impending Antarctic excursion attended a company briefing prior to their respective flights. Held on November 9, 1979, its initial portion was run by Captain John Wilson, supervisor of the Route Clearance Unit, who impressed on his auditors that the DC-10's computerized flight track would lead them straight up McMurdo Sound over flat sea ice at the end of the outbound journey. Off to the left would be Ross Island with its soaring active volcano, Mount Erebus, and off to the right would be the mountainous Antarctic mainland. The briefing proceeded at a leisurely pace as Captain Wilson covered the approach, the letdown, and the low flying in the McMurdo area. There was plenty of time for a question and answer session following an audio-visual screening.

After lunch, the airmen funneled into a DC-10 simulator conveniently located next door to the briefing room. Running this part of the extended training session was Captain Ross Johnson, flight manager (line operations). He explained the principal purpose of the simulator exercise as learning how to change the compass systems from magnetic to grid. Beyond sixty degrees latitude, grid navigation was needed because magnetic compass readings then become unreliable and longitudinal lines tighten markedly.

An in-depth pictorial review of salient Antarctic landmarks on approach to their outbound destination formed no part of the day's proceedings. Overlooked as well was the desirability of giving the airmen, for clarity and safety's sake, a topographical map with the Antarctic track accurately marked on it. Also oddly omitted from the November 9 briefing was any description of and training concerning the various forms of optical illusions—collectively termed *whiteout*—that regularly occur in the Arctic, northern Canada, and Antarctica. Most insidious of these was the phenomenon called sector whiteout, in which a pilot enjoyed excellent visibility in three directions of the compass but not the fourth. If the pilot experienced sector whiteout while traveling in the affected compass direction, he would believe himself to be flying in clear air over flat ground even if a formidable obstacle—say, a mountain—lay directly in his path.

Originally, it had been envisioned that no Air New Zealand pilot would be allowed to command a tourist flight to and from Antarctica without prior, supervised experience in the polar region. This was a long-standing requirement for American, New Zealand, and Australian military aircraft operating regularly in this potentially dangerous region of the world. A condition attached to Air New Zealand's Air Service Certificate No. 22 made it a requirement for the country's civilian aircraft as well. In deference to the busy schedules of the check and training captains and executive pilots who commanded most of the Antarctic charters, however, the national carrier quickly and quietly did away with this safety feature. Management claimed that a preflight briefing by the supervisor of the Route Clearance Unit, coupled with a stint in a DC-10 simulator, constituted adequate preparation for pilots ferrying tourists to the howling bottom of the world and back.

In a further safety lapse, Air New Zealand did not require even the Route Clearance Unit's supervisor to have any personal familiarity with Antarctica. That was good for Captain John Wilson, who possessed virtually none. He had admittedly made multiple attempts to get a seat on one of the Antarctic flights. Finally successful, Captain Wilson nonetheless never made it to McMurdo Sound because poor weather conditions there forced the DC-10 to divert to the South Magnetic Pole.

During the briefing session on September 9, Captain Wilson shared with his five attendees copies of TE901's current flight plan, which clearly gave the destination waypoint's coordinates as latitude 77 degrees, 53 minutes south and longitude 164 degrees, 48 minutes east. From comparing those coordinates on a map to other material shared with them, the airmen appreciated that the outbound terminus of the flight path, or nav track, would be some miles to the west of a beacon called the TACAN. That track would take them straight up McMurdo Sound to their outbound destination.

The flight path shown to the five pilots by Captain Wilson had been stored in the airline's ground computer for fully fourteen months before Captain Collins's Antarctic trip. This route up perfectly flat McMurdo Sound was eminently logical since it approximated that used by all American, New Zealand, and Australian military aircraft servicing McMurdo Station and Scott Base. It had two critical safety-related advantages besides the flat terrain. One was enabling VHF (very high frequency) radio contact with the Ice Tower, the US Navy's specialized radio and radar installation at McMurdo Station's ice landing field, once an aircraft came into line-of-sight range of it. Because on VHF voices were clearer than on HF (high frequency), the availability of this resource during the final 150 miles of an aircraft's outbound journey was invaluable. The other safety-related advantage of the McMurdo Sound—or "military"—route was enabling radar contact with that same Ice Tower, again on a line-of-sight basis, once an aircraft was forty miles away.

Air New Zealand's flight crews, briefing officers, and publicity personnel all knew that Captain Collins's nav track was programmed to run up the sound. So did American air traffic controllers at the Ice Tower as well as at Mac Center, the US Navy's main radio communications complex at McMurdo Station. On the day of Captain Collins's flight, everyone (including the pilots themselves) was thus expecting the DC-10 to approach via the customary route.

A fastidious pilot, Captain Collins spent the evening before his journey plotting the computerized flight track on a huge topographical map that he'd acquired for just this purpose. A cartography aficionado and qualified navigator from his time in the Royal New Zealand Air Force (RNZAF), Collins used the latitudinal and longitudinal coordinates that he'd been shown at the briefing session some days earlier to lay out the route. He knew that, in a marvel of aviation, the DC-10's sophisticated navigational computer achieved perfect accuracy by means of its three inertial sensor units, each of which functioned independently of the others. It was the redundancy of using three sensors, rather than merely one or two, that would enable his aircraft on a flight of over five thousand miles to be carried unerringly along the preprogrammed flight path all the way to its destination and back.

Kathryn and Elizabeth, the two oldest of his four daughters, were with Captain Collins in the dining room of their St. Heliers home as he laid out the track with his plotting instruments. So unwieldy was the mega-map (he was working with a small map as well) that, for convenience's sake, Captain Collins elected to sink from the dining room table onto the floor to complete his work. As he explained to the girls, about twenty-five miles to the west of his flight path would lie the scintillating mountaintops of Victoria Land and the mammoth glaciers descending from them, as well as such curious features as the Martian-like Dry Valleys. This so-called dry area mostly lacked water but managed to harbor primitive life forms in underground lakes lying below ice sixteen feet thick.

He would be hugging Victoria Land's shoreline, their father told Kathryn and Elizabeth. He did not think to mention what lay twenty-seven miles to the east of where he would be flying. Ross Island's Mount Erebus was Antarctica's second-tallest volcano, world famous for its hundreds of weirdly beautiful fumaroles (ice towers), unusual boiling lava lake, and strombolian eruptions (lava bombs). It was also memorable for the persistent plume of steam, gas, and particulates that streamed from its summit a further five thousand feet into the sky before being pushed sideways by the wind.

After the initial two Antarctic aerial tours in early 1977, passengers complained about the exceedingly high altitude at which they had been conducted. Why they complained is unclear, for they had no way of knowing that a workaround could possibly exist. One did, though, and it would beautifully enable Air New Zealand henceforth to give its airborne tourists what they wanted: to take close-up photographs of scenic sights on the ground. The carrier thus swiftly embraced a new practice of allowing its touring DC-10s to low fly up pancake-flat McMurdo Sound toward the outbound terminus of their flights at Hut Point Peninsula's Cape Armitage.

Shortly after the third of 1977's six Antarctic tours, accordingly, articles began appearing in the Auckland newspapers extolling the clear views of scientific research installations, penguin rookeries, and teams of huskies to be had from a low-flying Air New Zealand aircraft sightseeing in the most unwelcoming of climes. For example, the *Auckland Star* published a story on October 22, 1977, which featured a passenger on the polar aerial tour of four days earlier. The aircraft had reportedly descended "well below the towering volcano Erebus belching smoke only 50 kilometers [31 miles] away," and passed over McMurdo Station and Scott Base at a mere 200 meters [656 feet].

Not long afterward, *Traveling Times* carried a story on the Antarctic experience of John Brizindine, president of a division of the McDonnell Douglas Corporation (the DC-10's manufacturer), who had flown as Air New Zealand's guest on a sightseeing tour in November. In the piece, Brizindine enthusiastically described how the aircraft, commanded by the estimable Captain Gordon Vette, had zoomed by Ross Island and overflown Scott Base and McMurdo Station at a height of "perhaps half a mile." To capitalize on the dignitary's laudatory account of his exotic aerial adventure, Air New Zealand went to the expense of having a million copies of it reprinted as a brochure and distributed to every family in the country.

The practice was, with American air traffic control's explicit permission, to descend to somewhere between 3,000 and 1,000 feet—a range of altitude seemingly endorsed by the airline itself in its newsletter, *Air New Zealand News*. Distributed to all 8,700 Air New Zealand staff members, one newsletter featured an article dated November 30, 1978, that promoted a recent charter flight on which the airline's own director of flight operations, Captain Douglas Keesing, was aboard as a tourist. The article, which he had approved prior to its publication, begins by observing that "the flight deck crew of TE901 took the boss flying with them." As the DC10 cruised "at 2,000 feet past Antarctica's Mount Erebus and over the great ice plateau," Captain Doug Keesing was "as interested in sightseeing as the other 230 passengers on board."

Captain Collins planned to follow previous pilots by sweeping up broad, flat McMurdo Sound and ending the outbound section of his sightseeing tour by descending to within prime photography range of the ground under the guidance of local air traffic control. He would by then be approaching Hut Point Peninsula, a long and narrow extension of Ross Island on the tip of which were situated multiple attractions. Highly visible at low altitude, for instance, would be two noteworthy outposts of civilization, both scientific in nature: the Americans' McMurdo Station, with its hodgepodge of squat buildings and oil tanks, and the New Zealanders' Scott Base, with its unprepossessing small, green shelters. Easily visible too would be British explorer Robert Falcon Scott's Discovery Hut and the Terra Nova Memorial Cross honoring him and his fallen comrades, who had perished on their march back from the South Pole in January 1912. At 1,500 or 2,000 feet, even knee-high, kinetic Adélie penguins and four-foot-tall, regal Emperor penguins would be photographable, not to speak of sled dogs kept by the humans at the two research centers. Off in the distance, directly north of those installations, would loom smoke-emitting Mount Erebus. After turning around and taking a second low pass over the local sights, the DC-10 would commence the long journey north for refueling in Christchurch and the final push back to Auckland.

RACE TO THE POLE

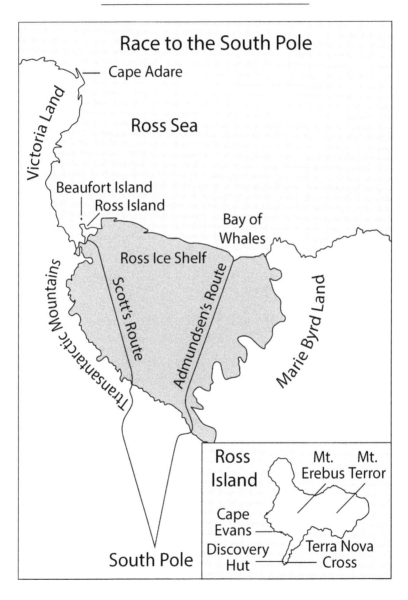

Race to the South Pole

Cape Adare

Victoria Land

Ross Sea

Beaufort Island
Ross Island

Bay of Whales

Ross Ice Shelf

Transantarctic Mountains

Scott's Route

Admundsen's Route

Marie Byrd Land

South Pole

Ross Island

Mt. Erebus   Mt. Terror

Cape Evans

Discovery Hut

Terra Nova Cross

SIGHTS OF ROSS ISLAND

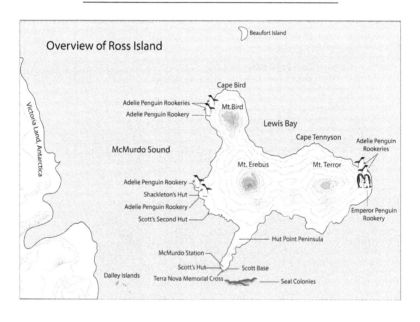

Overview of Ross Island

Beaufort Island

Victoria Land, Antarctica

Cape Bird

Adelie Penguin Rookeries
Adelie Penguin Rookery

Mt.Bird

Lewis Bay

McMurdo Sound

Cape Tennyson

Adelie Penguin
Rookeries

Mt. Erebus

Mt. Terror

Adelie Penguin Rookery
Shackleton's Hut
Adelie Penguin Rookery
Scott's Second Hut

Emperor Penguin
Rookery

Hut Point Peninsula

McMurdo Station
Scott's Hut
Terra Nova Memorial Cross

Scott Base

Dalley Islands

Seal Colonies

2

CAPTAIN JIM COLLINS'S FLIGHT

ON A BALMY, AUSTRAL SUMMER MORNING IN 1979, AIR New Zealand Flight 901—also known as TE901—lifted off at 8:17 a.m. from Auckland International Airport into clear skies carrying 237 excited passengers as well as twenty cabin crew, flight personnel, and a highly experienced Antarctic commentator. The aircraft was headed to the white continent on what had been touted as "the ultimate day trip" and "the experience of a lifetime." Sweeping over an offshore oil rig, the five-year-old, wide-bodied DC-10 gradually rose to cruising altitude. As the excursionists, clad in light clothing, enjoyed a champagne breakfast in the toasty cocoon of the cabin, the jet approached the flight's first scenic attraction: picture-perfect Mount Egmont (Mount Taranaki), a dormant stratovolcano on the North Island's west coast. From there it flew on toward the South Island's serrated ranges, the majestic Kaikouras and the snow-covered Southern Alps, the latter named by Captain Cook in 1770 and

described by him as "prodigiously" high. The aircraft's destination
lay a great distance beyond these spectacular geographical features,
on the far side of a vast and treacherous ocean separating New
Zealand from the distant continent of ice and rock at the bottom
of the world.

As the DC-10 Captain Collins was commanding hurtled along
at a clip of three hundred miles per hour, the mood in the cabin was
festive. All four flights announced for November 1979, of which
Captain Collins's was the last, had sold out within three weeks—
despite the not inconsiderable cost of a ticket sold with no guar-
antee the weather would permit a satisfactory aerial sightseeing
experience. Passengers were promised refreshments, two meals,
an open bar, inflight entertainment, and the novel sight from above
of the illimitable, still, frozen terrain famously explored by multiple intrepid spirits over the previous century and a half. In fact,
Collins's flight was taking place exactly fifty years to the day that
Richard E. Byrd Jr. had set out to make history by being the first to
reach the South Pole by air. To mark the historic anniversary, Air
New Zealand had commissioned commemorative certificates to
gift every sightseer on the touring DC-10.

Encouraged to move about the cabin in search of advantageous
views, the passengers mixed easily, chatting and joking during what
appears to have been a joyous airborne party. "Just fabulous," ex-
alted one passenger about an earlier Antarctic sightseeing trip. "I've
never seen anything like it in my life—it's beyond anything I ever
dreamed of!" Another gushed, "It was like a cocktail party that ran
all day." Air New Zealand billboards boasted that "no one does it
better," and at that point it certainly seemed to be true.

Unlike the merrymaking tourists, TE901's chief purser had
boarded the aircraft on November 28 with a serious behind-the-
scenes objective. For several months a senior inspector employed
by Air New Zealand's regulator had been pushing the airline to

carry survival gear on its Antarctic flights like Qantas Airways did. However, the airline had thus far resisted, advising its regulator just a month earlier that, in Air New Zealand's opinion, "the carriage of survival suits is unwarranted" since an emergency was "extremely unlikely." Armed with some Qantas material on the subject, the chief purser hoped to catch the attention of a receptive member of the aircrew sometime during the day-long adventure to stress the importance of this safety protocol. After all, everyone on board TE901 was dressed in summer clothes, with no possibility of surviving if exposed to Antarctica's cruel subzero temperatures, howling winds, and ferocious blizzards.

When the DC-10 passed high above the beacon at Invercargill, the country's southernmost city, the New Zealand portion of the outbound flight ended. By reference to that beacon, the pilots confirmed that their computerized flight track was operating perfectly. Some 450 miles further south, the aircraft overflew the Auckland Islands—a chain of small, uninhabited volcanic islands whose climate was uncharacteristically mild and humid, given the latitude. Almost a century and a half earlier, James Clark Ross's expedition had everywhere encountered vegetation: mosses, lichens, ferns, dozens of varieties of flowering plants. Antarctic flight captains overflying the Aucklands in the late 1970s found the view from above agreeable as well. After commanding his own aerial tour, Captain Gordon Vette described the chain's principal island as resembling a "rolling plateau of what looks like excellent grazing turf." Somewhere in nearby waters lay the 1,005-ton *General Grant*, yet to be located. Carrying gold bullion, it had foundered in 1866 en route to London from Melbourne. Soon the aircrew would be switching over to grid navigation and piercing the Antarctic Circle.

As the habitable latitudes edged away to the rear and Captain Collins and first officer Cassin pressed forward into wholly unfamiliar territory, an air of expectancy pervaded the passenger cabin. Eventually, the Cape Adare coastline of Antarctica came into view as the DC-10 cut across a corner of the continent on its way from the Balleny Islands waypoint to that at Cape Hallett. Approaching the latter at thirty-three thousand feet, the energized day-trippers could only gaze with wonder at "the vast glacier tongues, formed over millenniums, spilling off the Prince Albert Mountains to form solid rafts, the size of Switzerland, in the Ross Sea." The pilots confirmed one last time the total accuracy of their aircraft's navigation system as they overflew the flight's penultimate outbound waypoint at Cape Hallett's ice-free extremity, the site of a former US–New Zealand scientific base. From there, they knew, it would be a 389-mile run over the Ross Sea, with Victoria Land's mountaintops glittering on the right, and up the middle of flat-as-a-pancake McMurdo Sound to the destination waypoint at the Dailey Islands.

Polar sun suffused the aircraft as tourists from multiple countries leaned across temporarily vacated seats or pressed up against windows to catch on their cameras, in clear air, hundreds of photos of seemingly limitless frozen grandeur below. The clicking stopped temporarily while lunch was served. Dessert consisted of Peach Erebus, a light-hearted confection consisting of a piece of fruit cloaked in meringue and cream to represent the icy exterior of the "sentinel of McMurdo." Then, the cameras were again deployed.

In keeping with their promotion as a peerless aerial sightseeing experience, Air New Zealand's Antarctic flights always featured an experienced polar traveler serving as tour guide. Today's was supposed to have been Sir Edmund Hillary of Mount Everest fame. Owing to a scheduling conflict, however, he had needed to delegate the job to his "most staunch and willing supporter," Peter Mulgrew. Besides accompanying Hillary on the Commonwealth Trans-Antarctic Expedition of 1957–1958, the accomplished mountaineer had been with him on multiple climbing adventures and misadventures in Nepal. Mulgrew had a magnetic personality and was intimately familiar with Air New Zealand's "ultimate day trip." He had first served as Hillary's reserve commentator on an early Antarctic tour. This was his fourth flight as official commentator.

While he would soon move forward to the cockpit to resume his narration over the PA system of what the tourists were seeing below them, Mulgrew was presently in the cabin chatting animatedly with inquisitive passengers. To place in historical context what the day-trippers would be viewing as the aircraft approached its outbound terminus, Mulgrew had already shown three Antarctic-themed films. The first was titled *Amundsen: Explorer*. Its account of the legendary Norwegian would have included the cutthroat international race to the South Pole that he won handily on December 15, 1911, while the bumbling expedition of his competitor, British adventurer Robert Falcon Scott, was so hobbled by poor-performing motorized sledges and ponies that it was still 298 miles away.

An engaging story Mulgrew enjoyed sharing with passengers involved a sly episode that occurred during the Commonwealth Trans-Antarctic Expedition, in the New Zealand contingent of which he had served under Sir Edmund Hillary twenty years earlier. In accordance with the vision of its overall commander, Dr. Vivian Fuchs, the motorized expedition had aimed to conduct a variety of technical studies in the process of arduously crossing Antarctica from one end to the other via the South Pole. Against instructions, Hillary took off with his own team on an adventurous dash, which resulted in the New Zealanders being the first to reach the South Pole overland since Amundsen in December 1911—and the first to reach it in motor vehicles. They arrived sixteen days ahead of Fuchs.

When TE901 was around 140 miles out from its destination, the weather forecaster at the Americans' principal air traffic control complex at McMurdo Station, Mac Center, informed the crew of a "pretty extensive low overcast" in the local area. "Doesn't sound very promising, does it?" Captain Collins asked doubtfully. "No," first officer Cassin agreed. Since, however, visibility was reported to be good below two thousand feet, the pilots could still give the aircraft's passengers excellent views of the sights at the tip of Hut Point Peninsula by descending at the entrance to the sound and slipping *under* the low cloud cover.

Mac Center now radioed a welcome offer: "Within a range of 40 miles of McMurdo we have radar that will, if you desire, let you down to 1,500 feet on radar vectors." First officer Cassin had barely accepted the air traffic controller's offer on behalf of the flight crew when Captain Collins espied up ahead breaks in the cloud below and requested permission from Mac Center to descend in the clear air under visual flight rules. Permission was granted; all that was asked in return was for the pilots to maintain so-called visual mete-orological conditions (VMC) and keep American air traffic control apprised of the aircraft's altitude.

For the remainder of the flight, Captain Collins punctiliously observed visual flight rules. He deliberately continued flying on the computerized navigation track as well. Air New Zealand's Captain Gordon Vette would later characterize this technique as "flying with belt and braces caution."

At forty-three miles north of the destination waypoint, Captain Collins decided to update the passengers about the crew's plans for the final portion of the outbound journey. "Ladies and gentlemen," he began, "we're carrying out an orbit and circling our present position, and we'll be descending to an altitude below cloud so

rving as flight commentator on Air New Zealand's aerial tour
October 15, 1977, for instance, he deeply impressed Captain
rdon Vette, its commander, with his abilities in this area. "I did
t have to refer to any of my topographical charts in order to fix
aircraft's position," Vette would later marvel.

"There you go! There's some land ahead," Mulgrew enthused,
ing the airliner to be entering into the clear, broad expanse of
McMurdo Sound under the overcast, with the dark cliffs of the
inland's Cape Bernacchi to the southwest and those of Ross
nd's treacherous Cape Royds to the southeast. Forward and
the left, the commentator pointed out Cape Bird—familiar to
previous tourist flights—which gave Captain Collins
independent fix on his position. If he then consulted his map,
ch is likely, Captain Collins would have seen Cape Bird forward
to his left, thus confirming the experienced commentator's
ntification.

Having confidently taken the headlands to the right and left
he DC-10 to be those at the approach to McMurdo Sound,
lgrew declared minutes later, "Taylor on the right now." The
nland's Taylor Valley was one of three Dry Valleys to the west
McMurdo Sound, in the Transantarctic Mountains. "Where
It. Erebus in relation to us at the moment?" engineer Brooks
ired a minute afterward, obviously thinking about any high
ain in the area. "Left, about twenty to twenty-five miles,"
grew answered, "about eleven o'clock." Papers rustled as some-
on the flight deck confirmed this on Captain Collin's large
graphical map.

"That's the edge," Mulgrew now announced, meaning the ver-
ous edge of Ross Island's Cape Royds. If Ross Island was off to
eft, the DC-10 was about to fly past it safely at low altitude over
en McMurdo Sound. "I reckon Bird's through here and Ross
d there," the commentator added. "Erebus should be here."

that we can proceed to McMurdo Sound." After co:
first leisurely descending orbit in place, conducted i:
Captain Collins executed a second one. Carefully !
onto the computerized track after completing each o:
loops, Collins proceeded gradually to bring the airc:
ward two thousand feet, as he had advised Mac Cente
would then fly itself, at perfect sightseeing altitude,
over flat sea ice to the Dailey Islands waypoint, as in
pilots' briefing nineteen days earlier.

While the aircraft shed height, Captain Collin:
future track he'd be following at a lower level and
obstructions. As a trained air navigator, he was accus
uating geographical features in geometrical terms c
angles, in relation to each other. These need to be i:
alignment or, as Erebus authority Stuart Macfarlane
"either you are not looking at what you think you are
located where you think you are, or both." Momen
Collins would confirm that he *was* where he was s
Two capes hove into view at precisely the azimuths
ing for Cape Bernacchi on the Antarctic mainlan(
and Ross Island's Cape Royds (on the left). The viev
of TE901 as it descended also proved reassuring.
fectly flat white plateau stretching away into the di:
what he was anticipating for McMurdo Sound.

As the DC-10 glided downward, commentato:
had slowly been making his way forward from the '
to the flight deck on prosthetic legs, the conse(
monary embolism suffered on Mount Makalu in
himself in the jump seat behind Captain Collins,
for some topographical feature, or landmark, to a:
public address system. Mulgrew's skill at identify
in the vast white wilderness that is Antarctica '

Throughout this crucial last piece of TE901's outbound flight, the pilots experienced repeated trouble communicating with the Ice Tower via short-range, exceptionally clear VHF (very high frequency) radio, although not-so-clear HF (high frequency) radio contact with Mac Center did prove possible. Additionally, try as they might, they could not lock on with McMurdo's TACAN radio beam (a military navigational aid). They knew that, unlike HF transmissions, VHF and TACAN beams travel on a direct line of sight. Now in McMurdo Sound and closing in on the outbound flight's destination waypoint, Captain Collins and first officer Cassin were puzzled why communication with the Ice Tower on VHF still was not working.

After reaching 2,000 feet, Captain Collins soon decided to drop down a tad more, to 1,500 feet. His hope was that this would enable the pilots to see further up the sound—to espy the two scientific research stations forward and to the left as well as Mount Discovery straight ahead in the far distance. Mount Discovery was a conspicuous stratovolcano lying at the head of McMurdo Sound, east of Koettlitz Glacier and beyond the Dailey Islands waypoint. It would be impossible to miss flying in clear air.

But nothing was discernible—not Mount Discovery in the distance, not the scientific installations and jumble of huts, penguins, and huskies the air crew knew were located nearby on the tip of Hut Point Peninsula, the flight's ultimate sightseeing objective. Straight ahead there was only an endless vista of perfectly flat, white, featureless ground. "Have we got them on the [Ice] Tower?" Captain Collins asked his copilot. "No, I'll try again," Cassin replied. Things weren't quite adding up.

"I don't like this," flight engineer Brooks now announced. In Captain Gordon Vette's expert opinion, Brooks, whose seat in the cockpit was behind the pilots, must have just lost sight of the dark cliffs to the right and left of the DC-10. Everything in his view now would suddenly have turned uniformly white, which strange perceptual experience has been likened to being inside a big milk bottle.

From their seats in front of Brooks, Captain Collins and first officer Cassin could still see the textured cliffs on each side of the aircraft. Once those geographical features slipped away to the rear, though, they would enter the milk bottle with Brooks. In the meantime, the pilots were acutely aware that despite being so close to the destination waypoint in McMurdo Sound, they still could not reach the Ice Tower.

What aviation expert Gordon Vette would later call "conflicting inputs"—no Mount Discovery, no research outposts, no VHF radio or radar contact when they were in close to the Ice Tower and in its direct line of sight—were combining to produce in Captain Collins a distinct feeling of unease. Well known for his prudence, he decided to fly away. "We're twenty-six miles north," Collins announced immediately after Brooks had expressed his misgivings. "We'll have to climb out of this." There then began an unhurried discussion between the captain and his copilot as to which way to turn, left or right, to climb away. It did not really matter since McMurdo Sound was wide enough to accommodate the aircraft in either direction, but a proposed change in plan required discussion. A decision had yet to be reached when the ground proximity warning device abruptly sounded.

Whoop . . . Whoop . . . Pull up . . . Whoop . . . Whoop . . .

"Five hundred feet," announced flight engineer Brooks, reading off the aircraft's distance from terrain.

Pull up . . .

"Four hundred feet."

Whoop . . . Whoop . . . Pull up . . . Whoop . . . Whoop . . .

"Go-around power, please," Captain Collins requested, to muster an emergency burst of power from the engines as, simultaneously, he pulled back on the control column to get the DC-10's nose up fast. Although having no reason to think the aircraft was in danger, Captain Collins had prudently summoned go-around power, as it was standard operating procedure whenever a ground proximity warning device went off. The phrase derives from those situations in which a pilot needs to abort a landing, gain elevation, go around the airport, and try again to land safely. If Captain Collins had believed that he faced a dire emergency, by contrast, he would have called for the engines to be thrust to *maximum* power.

Pull up . . .

Trailing TE901 south from Christchurch to McMurdo, nearly 2,500 miles away, was a US Military Air Command Lockheed C-141 Starlifter. It carried, among other prominent passengers, the nephew and grandson of Richard E. Byrd Jr.—naval officer, polar explorer, pioneering aviator, and descendant of one of the First Families of Virginia. His nephew, career public servant Senator Harry F. Byrd Jr., remembered speaking by telephone to his famous relative as a fourteen-year-old moments after the aviator completed humankind's first harrowing flight to the South Pole. Now, at the invitation of the National Science Foundation, the senator from Virginia and other dignitaries were headed to the ice landing field associated with McMurdo Station.

The next day, November 29, would mark fifty years since the senior Byrd had completed his epic flight. After eighteen and one-half hours in the air, he safely arrived back at his Antarctic base in a Ford trimotor with three companions. Senator Byrd, grandson Robert Byrd Breyer, two members of the original ground party, and select National Science Foundation officials planned to participate in special events celebrating a half century of exemplary Antarctic aviation. The senator further planned to be in a ski-equipped Hercules on the commemorative flight scheduled, as part of the festivities, to retrace the senior Byrd's route to the Pole and back to his Antarctic base.

Because the massive turbofan freighter was only forty or so minutes behind TE901 and flying practically the same route, the two aircraft were periodically in touch as they traveled over the Southern Ocean toward Antarctica. As Major Bruce Gumble, who was captaining the C-141 Starlifter, neared McMurdo Sound, he heard the DC-10 inform Mac Center it was making a visual descent to two thousand feet. A short while later, Major Gumble tried unsuccessfully to contact TE901 to ascertain its exact approach to McMurdo. Beginning his own descent twenty minutes later, he attempted, unavailingly, a second time to raise the other airliner's crew. "There was getting to be some suspicion that something had gone wrong," Major Gumble would recall late that night. "So we took a look as we went in; we saw nothing."

When those aboard the C-141 Starlifter emerged from the aircraft once it had landed on McMurdo's ice runway, they stepped not into a celebration of a half century of triumphant Antarctic flight but rather into an unimaginable aviation crisis. TE901 had not flown up McMurdo Sound toward McMurdo Station at low altitude as anticipated. Mac Center had not been able to reach TE901. The airliner, which was required to report to Mac Center at thirty-minute intervals, was not reporting.

As Senator Byrd and the other dazed passengers were invited to relax in a chalet belonging to the National Science Foundation, the C-141 Starlifter turned around and headed back north in the direction of Christchurch, its crew keeping a sharp eye out for the DC-10 along the route it had taken to its last reported position. Dressed in summer clothing, 237 tourists together with twenty cabin staff, air crew, and an Antarctic commentator had disappeared in one of the planet's harshest and most inaccessible regions. It now seemed unlikely that Senator Byrd would be making a commemorative flight the next day in honor of his uncle's aviation feat fifty years earlier. The Hercules would doubtless need to be reassigned to help look for the missing DC-10.

As Major Gumble flew back over what its crew knew of TE901's route, Mac Center placed additional aircraft on standby for a full-scale search and rescue mission. It also notified Auckland of the missing aircraft.

OUTBOUND FLIGHT PATH

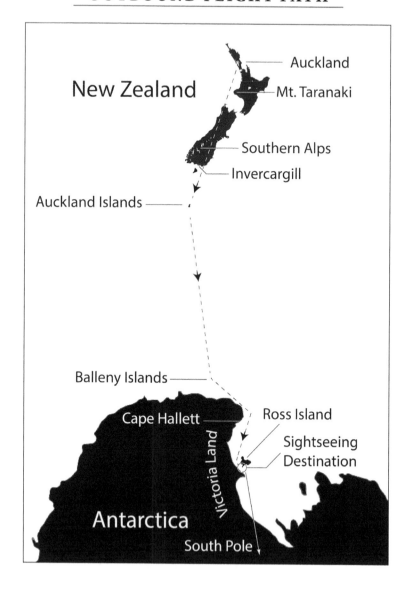

APPROACH TO MCMURDO SOUND

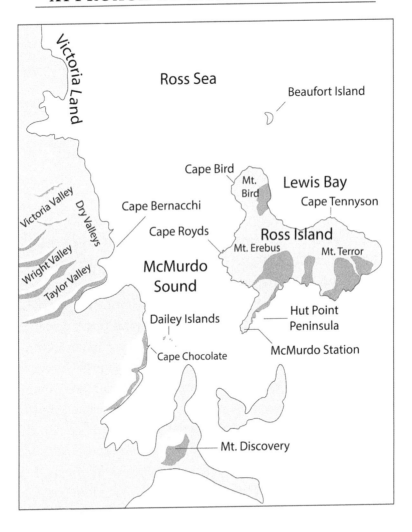

3

VANISHED AIRCRAFT

ARRIVING AT 3:00 P.M. ON WEDNESDAY, NOVEMBER 28, for his shift at Flight Dispatch, Alan Dorday heard the unsettling news that TE901 had been out of contact for over an hour in one of the planet's most unforgiving environments. Located at Auckland International Airport, Flight Dispatch was a pilot's last stop before boarding his assigned aircraft. It was where a captain headed for the white continent would receive his Antarctic envelope, which contained multiple documents germane to the aircrew's imminent journey. The most important of these was a printed copy of the flight plan stored in Air New Zealand's ground computer. The printout consisted of a prodigiously long series of coordinates and headings (247 scattered digits) that, once aboard the aircraft, two members of the flight's crew would manually input into the flight deck computer.

That evening, as collective anxiety at the national carrier mounted, Alan Dorday and his supervisor, David Greenwood, were visited by two aviation accident investigators representing, in the one case, the New Zealand Air Line Pilots' Association (NZALPA) and, in the other, the government's Office of Air Accidents Investigation, which was part of its Ministry of Transport. While the investigators were at Flight Dispatch collecting intelligence and documents from Greenwood, Dorday was manning the phones and talking to an increasing horde of visitors. Things were becoming hectic in what had morphed into an informal command center preoccupied with the mysterious disappearance of TE901.

Then navigation systems specialist Keith Amies phoned with an unusual question: Were all the longitudes for TE901's flight plan easterly? He wanted Dorday to check because, as Amies explained, the ground computer's Antarctic flight plan had recently been updated. While Amies waited on the line, Dorday scrutinized all the waypoints. Indeed, he reported, they were easterly—every one of them.

His curiosity piqued by this brief exchange, Dorday pulled up Captain Collins's flight plan as well as Captain White's from the previous week's sightseeing tour. Comparing them, he noticed a discrepancy involving the location of the destination waypoint. "The difference amounted to some two degrees of longitude," he would state in a detailed account written two days later. On a flying leg of several hundred miles southward from Cape Hallett, the penultimate waypoint, to McMurdo, two degrees of longitudinal variance would initially be negligible. However, as Dorday knew, the distance between the one flight plan's track and the other would gradually widen east to west. Concerned, Dorday now pulled up all four flight plans for the current season's Antarctic sightseeing excursions and closely compared them one to another. They were identical—except for Captain Jim Collins's.

After Dorday shared his bombshell discovery with Greenwood, the two plotted the course of the last leg of Captain Collins's altered flight path. Unbelievably, it led not up pancake-flat McMurdo

Sound but over Ross Island on a collision course with a menacing volcano. "As a result of these findings, and considering the overdue situation of TE901," Dorday would subsequently recollect, "we both had fears . . . that the aircraft may have collided with Mt. Erebus." They contacted navigation systems specialist Amies, whose reaction was one of shock.

The hour was now late. At Amies's urgent behest, Dorday immediately phoned the airline's chief navigator, Brian Hewitt, at his home. After identifying himself, he began urgently to speak. A very ill-tempered Hewitt cut Dorday off. Yes, he had made a change in the destination waypoint's coordinates of some two miles (ten minutes of longitude), but that was no business of anyone at Flight Dispatch. Taken aback as he was by Hewitt's abruptness, Dorday found himself surrounded by inquisitive parties and thought it "imprudent to talk of an accident or possible accident," or even to elaborate on the discrepancy he had found. He therefore contented himself with this bland inquiry: Did the chief navigator understand what he, Dorday, was talking about? Hewitt said he *did*—and reiterated that "the discrepancy was of no concern" to those at Flight Dispatch.

This peremptory rebuff caused Dorday to wonder whether Captain Collins and first officer Cassin had ever been briefed on the revised flight path. In reality, it involved a shift not of a smidge over two miles (ten minutes of one degree of longitude) but of some twenty-seven miles (slightly over two degrees of longitude). Checking the records, he found that there had been no official briefing for an Antarctic flight after the one Captain Collins and his copilot had attended with three other airmen nearly three weeks earlier.

Dorday was aware that Air New Zealand's ground computer was typically updated very early on Wednesday mornings. The alarming coordinate change he had uncovered must thus have been made *after* the aerial tour of Wednesday, November 21, which proceeded normally and uneventfully up McMurdo Sound, but before the flight of Wednesday, November 28, which was Captain Collins's currently missing one. This meant that Captain Collins and his crew must have departed for the ice a mere matter of hours after the computer's most recent update. Did the revised flight plan have attached to it an ops flash (a special type of notation) to alert the pilots to the altered route? Because he did not know, Dorday's apprehension was steadily increasing. It didn't ease when he went home at 12:30 a.m. and pulled out his family's volumes of *Scott's Last Expedition*, which contains forbidding photos of Mount Erebus.

Then he heard the news.

In 1979, November 28 had begun uneventfully at Scott Base and McMurdo Station, the two research installations located on the tip of Ross Island's Hut Point Peninsula. Information officer and photographer Nigel Roberts, who was employed by the New Zealand operation, listened along with others in the lunchroom to communications between the Williams Field control tower and an approaching sightseeing aircraft. Indirectly, Roberts owed his presence in the mess that day to a "stunning lecture" he had heard as a teenager by Sir Vivian Fuchs, who had been knighted after leading a twelve-man party across the white continent during the arduous British Commonwealth Trans-Antarctic Expedition of 1957–1958. Today, in the lunchroom, Roberts's interest in the jetliner's progress was especially keen because he was determined to take a close-up photograph of TE901 overflying Scott Base, his intention being to sell it to Air New Zealand for promotional use.

Suddenly the broadcast stopped, but no one listening to it was concerned; communications were not reliable in the coldest, highest, windiest, and driest continent on earth. As the afternoon wore on, though, there oddly continued to be no news of the touring aircraft. Roberts himself sensed something was clearly wrong when he saw Scott Base's leader, normally "unflappable," run out, leap into a Land Rover, and race over the hill to nearby McMurdo Station.

If the wind isn't howling, nighttime in and around Scott Base tends to be eerily quiet, for there are no leaves to rustle. But on that night, Roberts would later recall, a jarring contrast to normality was provided by the din of ski-equipped Hercules "warming up, taking off, flying, coming back, landing in a constant search for the plane." Once it was located, information officer and photographer Nigel Roberts would be dispatched in a helicopter to fly over the crash site and shoot an entirely different sort of photograph of TE901 than he had originally planned.

In the early evening of November 28, Justice Peter Mahon of New Zealand's High Court was driving home from work when he heard a radio announcement that an Air New Zealand DC-10 was missing in Antarctica. Details were sketchy. "I continued listening at home," the judge would later recall, "and it was soon apparent that deep anxieties were felt as to the safety of the aircraft and those on board."

For governmental personnel in Wellington, the nation's capital, the deep anxieties Justice Mahon had detected from radio reports in Auckland were turning into alarm bells. By chance, Radio New Zealand's political editor was at a local bar with an acquaintance who worked under Prime Minister Muldoon's chief press secretary. Around 7:30 p.m., the acquaintance was urgently summoned back to the Beehive, which houses the offices of the prime minister and other top officials. The lights there were reportedly still on at 10:00 p.m., indicating something big was taking place, even though the prime minister himself was out of town.

When the US Navy four-engine Starlifter that had carried the American senator Harry Byrd and other luminaries to Antarctica behind TE901 arrived back in Christchurch around 10:00 p.m., its captain found himself besieged by reporters. On the flight north over McMurdo Sound, Major Gumble told the crowd, he'd made a point of taking a close look as he passed hazardous Cape Royds on Ross Island's western coast, descending to 2,000 feet and then to 1,500 feet in hopes of locating the vanished aircraft. Nothing had come, however, of his attempt to find the DC-10 in the area where the initial search parties were looking owing to its presumed flight path.

In the arrival lounges of both Christchurch International Airport (around fifty passengers were expected to disembark there) and Auckland International Airport (through which the rest would pass), relatives of the missing sightseers anxiously awaited word of TE901's whereabouts. The time came and went when the DC-10 would have exhausted its fuel. Deficiencies in the communication apparatus at the Christchurch tower, revealed a month earlier in connection with an Antarctic-originating aviation crisis involving a crippled US Air Force Starlifter, now hindered local air traffic controllers from obtaining timely updates on what was happening well over 2,000 miles away. Their only sources of information were second- and thirdhand reports, either relayed to the Americans' nearby Deep Freeze base from McMurdo and then forwarded to them or sent by Auckland's Oceanic Control Center, which oversaw the New Zealand end of what had turned in a matter of hours from a standby operation to a vigorous Antarctic search and rescue mission. The DC-10 appeared to have vanished into thin air.

Air New Zealand CEO Morrie Davis had spent Wednesday, November 28, in Heretaunga, Upper Hutt, schmoozing at the historic Royal Wellington Golf Club, 1979's venue for the Air New Zealand–Shell Open. The premier sporting event had been his brainchild, and the airline he oversaw served as cosponsor. It being pro-am day, Air New Zealand's golf-loving chief executive had been paired with the accomplished Australian player David Graham. The two were still on the golf course when Davis was discreetly handed a message that upended his immediate plans and sent him swiftly back to Auckland on a specially arranged, after-hours Boeing flight.

Upon arriving at the twenty-story Air New Zealand House in downtown Auckland near the wharves, Davis was immediately confronted by reporters seeking his reaction to the airline's presumed first commercial loss. "That's a question that makes me sick," Air New Zealand's irascible CEO groused. "We are terrified— these are the kind [sic] of circumstances that chill us all."

The tension and uncertainty mounted as American and New Zealand search aircraft failed to locate TE901 anywhere along the computerized navigation track established in mid-1978, carried on the previous seven DC-10s flying the Antarctic route and believed to be on this eighth one as well. Hours passed. The search area was broadened to include Ross Island. Finally, in the bright polar light of the austral night in Antarctica, where the sun circles the Pole twenty-four hours a day, Commander Victor Pesces and the crew of his ski-equipped C-130 Hercules spotted a suspicious black smear. It was in an unlikely place: the lowermost northern slopes of Ross Island's Mount Erebus, about a mile from Lewis Bay's coastline.

Unable to land owing to weather conditions, the airmen radioed their supervisors to send a helicopter to their position to take a closer look at what might be the remains of TE901. Its crew was able to identify the wreckage as TE901 by the distinctive white Maori koru emblem from the aircraft's largely intact, blue-and-turquoise tail section. It was lying about halfway up a long black gash in an icy slope dotted with pieces of aircraft rubble. At 1:25 a.m. on November 29, 1979, the crew radioed the terrible news: "Debris at crash site being blown by wind. No apparent survivors." Word was sent back to Auckland, prompting Air New Zealand's grim-faced CEO Morrie Davis to summon journalists and photographers to the boardroom adjacent to his nineteenth-floor office overlooking Waitemata Harbor. Sitting at the boardroom table, hands clasped, he gazed out from behind heavy, black-framed glasses. "I wanted to tell you that wreckage has been sighted in Antarctica," he funereally began. "The location indicates Mt. Erebus."

Despite the burden of having just learned that TE901 had crashed, Morrie Davis instructed the Air New Zealand–Shell Open's organizers and players to proceed with the event as a personal favor to him. Overnight the airline removed its ubiquitous sponsorship billboards, which featured the slogan "nobody does it better," and had the Royal Wellington Golf Club lower its flags to half-mast. "We felt so bad," David Graham would recall years later, about "playing golf in the shadows of something horrific." They did the best they could under the gloomy circumstances.

Graham would wind up winning the Air New Zealand–Shell Open by eight strokes. In his first public appearance since the aircraft's loss, CEO Morrie Davis attended the golf tournament's prize-giving ceremony. Thanking everyone for soldiering on, he predicted, "The dark cloud of tragedy will be with us all for some time to come."

The gruesome fate of the aircraft was discovered in time to make the papers worldwide on November 29, 1979. Like everyone else, the High Court's Justice Mahon had trouble guessing what could possibly account for TE901's horrifying transformation into a black smear of debris extending for hundreds of yards up the white, icy flank of an active Antarctic volcano. Unlike everyone else, however, he would soon be offered the opportunity to find out.

DORDAY'S DISCOVERY

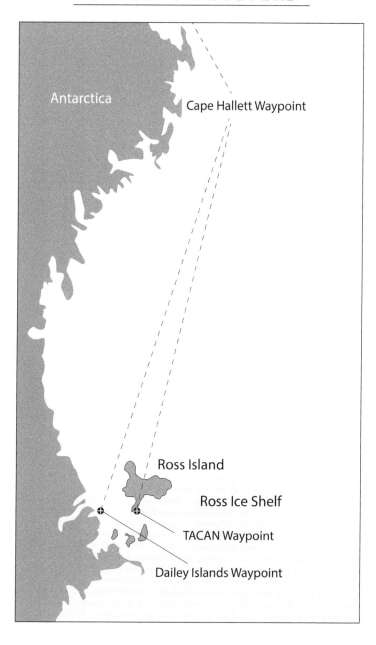

EXPECTED VS. ACTUAL FLIGHT PATH

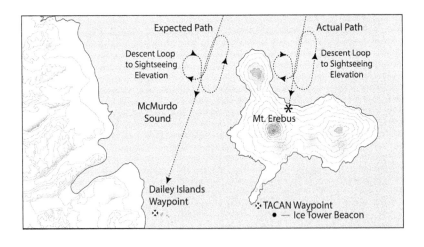

Expected Path

Descent Loop
to Sightseeing
Elevation

McMurdo
Sound

Dailey Islands
Waypoint

Actual Path

Descent Loop
to Sightseeing
Elevation

Mt. Erebus

TACAN Waypoint
● — Ice Tower Beacon

4

VANISHED SOULS

AROUND 7:00 P.M. THE DAY OF HER HUSBAND'S Antarctic flight, Captain Collins's wife, Maria, received a phone call from Air New Zealand's director of flight operations, Captain Dave Eden. After explaining that the pilots were not reporting in at regular intervals as required, Captain Eden suggested the forty-four-year-old mother of four invite another adult over to sit with her. Maria experienced a shiver of fear. It was her husband's practice to call at the end of a flight and say, "I'm at the airport now, and I'll be home in half an hour." Jim hadn't yet called, but it was still early. Why would someone from the airline ring and say it might be good to get "another adult" to be with her? "I was really shocked at that," Maria would later recollect.

When the 9:00 p.m. news came on, it was announced to an anxious nation that the missing DC-10 had by now run out of fuel. "We couldn't believe what we were hearing," Elizabeth Collins, then fourteen, would remember. Her mother's response was to phone Jim Collins's close friend and colleague, Air New Zealand pilot Ross Gordon, who dashed to her home with his wife to offer desperately needed support. Kathryn Collins, then fifteen, would never forget her musings that terrible night:

> If the plane had landed on the icy sea, Dad would be the last one off. I pictured him stepping off the wing onto an iceberg but being very cold as he was only in summer uniform. He would be very worried about his passengers. He would put them first, before him.

It was hours later, perhaps 1:30 or 2:00 a.m., when Captain Eden finally phoned Maria a second time to report both that aircraft wreckage had been discovered on the flanks of Mount Erebus— and that the damage was such there could be no survivors. Under these horrific circumstances, the stricken widow was relieved her husband had perished. Had he lived, she knew, Jim Collins would have assumed total responsibility for the disaster and become a "haunted human being forever." She wondered what could possibly have happened, though, for an aircraft flown by a pilot as skilled and prudent as her husband to come to such an abrupt, disastrous end. The accident didn't make any sense.

Already in these initial interactions with a representative of Air New Zealand, Maria discerned a distinct lack of compassion, or soul, on the airline's part. "I always expected that if the worst happened, there would be a knock on the door, and some executive would be standing there," she would say later. Instead, at her life's most devastating moment, Maria received a perfunctory phone call informing her of the loss of the DC-10, of her spouse, and of all those in his care. "I don't even know who it was that rang."

The solicitous Captain Ross Gordon, who had been with the Collins family as its members awaited word of TE901's fate on the terrible evening of November 28, returned the next day to help his good friend's widow in whatever practical ways might seem most urgent. During his visit, he learned from Maria that her husband had been working with charts and plotting instruments the night before the accident flight.

As fastidious as he was, Captain Collins might have been expected to plot his route to the flight's outbound terminus and back on a privately acquired topographical map. To do so, however, he would have needed to have a copy of the very long list of coordinates for all the waypoints involved. Where could he have obtained such a document? The only conceivable venue would have been at the November 9 briefing session, during which Captain John Wilson had circulated multiple copies of the flight plan. Given later developments, it seems likely that Captain Wilson's position within the briefing room enabled him to spot Captain Collins retrieving one of these to take home.

Only after the crash did the devastating implications of that leniency become obvious to the airline. The November 9 flight plan described a track up McMurdo Sound to a destination waypoint in the Dailey Islands. But Air New Zealand's top executives now knew that Captain Collins's computerized navigation track had been reprogrammed just hours before his flight. The revised flight path led over Ross Island on a beeline for Mount Erebus.

This awkward situation must have prompted Captain Wilson to share what he knew with the airline's director of flight operations because, in an odd incident occurring two days after the accident, Captain Eden personally contacted Collins family friend Captain Gordon. He wanted to know whether Gordon had seen at his dead colleague's house a "missing" copy of the flight plan from the November 9 briefing session. Gordon replied that he had not spotted a flight plan (it no doubt had traveled to Antarctica in Jim Collins's flight bag). The ruse was clever in that without raising any suspicions in Captain Gordon's mind about his real purpose in calling, Captain Eden gained invaluable intelligence. There was in the Collins residence no smoking gun in the form of a copy of TE901's flight path as of November 9. That was important because such a document had the potential to destroy the case the airline was already building against the pilots.

Captain Jim Collins's close friend Captain Barney Wyatt visited Maria several times right after the disaster. He confirmed her suspicion that management considered her husband and his copilot culpable but then added, "They haven't told you about the errors made, have they? The company is going to come in for some severe criticism." The new widow was totally baffled. On his last visit, Wyatt startled Maria by announcing that he would not be calling on her going forward; he needed to stand with the company. *Stand with the company? Against what—or whom?*

Air New Zealand's cover-up of its computer programming error might be said to have begun the evening of November 28, 1979, when Flight Dispatch's Dorday and Greenwood failed to inform the two investigating accident inspectors of their mind-boggling discovery that Captain Collins's flight plan had been altered at the last minute. Thirty years later, Greenwood would declare that while silent the night of their discovery, he "was in the Nav Section the following morning" and "certainly told Ian Gemmell." Unable to sleep, he had arisen early and gone to Air New Zealand House. Arriving at the seventeenth floor, where the Navigation Section was located, he said he was "pretty concerned about what had happened" and was "pretty sure Keith Amies and the navs would be there." Indeed, "probably all the navs were there." One might be forgiven for wondering why *all* the navs would convene at Air New

Zealand House at 6:30 a.m. the day after one of the airline's most meticulous pilots had flown his DC-10 straight into a mountain at the frozen ends of the earth, instantly killing all on board. Did they suspect a screwup in their own section?

Captain Ian Gemmell, with whom Greenwood reportedly shared the two-degree longitudinal coordinate change, was the behind-the-scenes architect of the airline's Antarctic flights. Later that day he would be departing for Antarctica as the newly appointed technical advisor of Ron Chippindale, chief inspector of air accidents. Per statutory regulations, Chippindale's team was responsible for uncovering the circumstances surrounding TE901's crash. While it was certainly true that no one on earth knew more about New Zealand's polar excursions than the man who designed them, it seemed to some an odd appointment for Chief Inspector Chippindale to make, given that it was precisely Gemmell's own protocols and practices that presumably were to be subjected to rigorous scrutiny.

The life to date of Captain Gemmell, who in 1975 became Air New Zealand's chief pilot and in 1978 its flight manager (technical), was one of adversity overcome through sheer willpower. Attracted early on to aviation, a costly pursuit better suited to rich young men than to those without means, the determined would-be pilot joined a forerunner to Air New Zealand as an apprentice fitter, a basic, practical job that involved working on the engines of flying boats and sweeping out the hangar. Remarkably, the fitter was later able to transfer to the pilot ranks, which, as his son would later note, "was a major breakthrough for Ian because previously all aircrew had come through the Air Force." Believing that anyone could achieve his ambition "by dedication and hard work," he proved it was so by rising indefinitely, a true rags-to-riches story.

According to his son, "Ian never did things by halves." He was stubborn, and that stubbornness never left him. One of the most consequential characters in the entire Erebus drama was, in short, a strong-minded, self-made man with a lifelong obsession with aviation.

Greenwood's reason for approaching Chief Inspector Chippindale's new technical assistant, Captain Gemmell, on the seventeenth floor was "to make sure that he knew [about the coordinate change] before he went down [to the ice]." For the record, Captain Gemmell himself always insisted he'd not been informed of the coordinate change until *after* he returned from Antarctica. If this was a falsehood—he repeatedly swore that it was not—his reason for telling it could only have been to lessen suspicion that he accompanied the chief inspector to the crash site with preconceptions about what he wanted to accomplish there. Chippindale himself was not informed about the coordinate change until later.

While shaken members of the Navigation Section compared notes "the morning after," three New Zealand mountaineers from Scott Base were lowered from a US Navy helicopter onto the sloping crash site to reconnoiter. This was a dangerous maneuver owing both to gusting winds and the crevassed, slanting terrain. From the air, helicopter crew chief Joe Madrid could see corpses "strewn about like they were straws." Once the mountaineers themselves were safely on the ground, they faced a hellish scene straight out of Dante's Ninth Circle. Two hundred tons of intricately assembled titanic machinery carrying a load of fifty tons of humans and fuel had disintegrated as the aircraft's three powerful engines propelled it with tremendous force from the point of impact up a fourteen-degree incline. Passengers, cabin crew, and aircrew were intermixed with the debris, which was smeared for nearly two thousand feet from bottom to top, where the deformed and blackened fuselage, consumed by a massive fireball, had finally come to rest.

Mocking the gruesome scene of mass carnage was Air New Zealand's tailpiece, which lay intact and exposed in the snow. Its iconic logo, specially created by a design team for the airline in 1972, consisted of a stylized version of the spiral design of an important Maori cultural symbol known as the koru. Based on the shape of unfurling new shoots in a silver fern frond, the koro symbol suggests creation, growth, movement, renewal—in a word, *life*. Nothing was further from being on the desecrated mountainside than that.

"Broken bits of everything, including broken bits of bodies," mountaineer Hugh Logan would subsequently recall, were literally "everywhere." The sudden deceleration from three hundred miles an hour to zero had horrifyingly disfigured the now-frozen flesh of many of those not instantly vaporized. Either that flesh no longer looked human or it was grotesquely disfigured or partially disintegrated. Frozen limbs were twisted, bent at bizarre angles. Already the ceaseless din of squawking skua gulls, the so-called raptors of the south, filled the air as these ferocious scavenger birds swooped down from the skies to dine on the dead. The air positively reeked of kerosene (the DC-10's fuel tanks were over half full when the aircraft hit the mountain), and the antifreeze in the jet fuel reportedly had an unpleasant effect on the bodies' odor. The entire desolate scene was one of hellish destruction and death, of both man and machine.

Once Madrid retrieved the trio, who were exhausted by their intense if short ordeal on the mountain, he flew them back to McMurdo for debriefing. There the men were shocked to see "tables set up in the gym with food, coffee, water, and all manner of foods." "Of course this was for the crash survivors," the helicopter crew chief instantly realized. By then, though, "we knew [they] would not be arriving, yet there were these tables, filled with all the hope of life."

There was no life.

5

RETRIEVED BODIES

AT AGE TWENTY-TWO, CONSTABLE STUART LEIGHTON
was the youngest member of New Zealand's disaster victim identification (DVI) squad. As a member of a Wellington-based DVI team,
he had just attended at Police National Headquarters a refresher
course on the handling of disaster victims, precipitated by the recent
massive loss of life in a Chicago DC-10 crash. The meeting ended
around noon on November 28, 1979, after which Leighton and two
colleagues drove back together to their local police station. "You
know," the senior constable remarked, "I have been in the police for
over twenty-five years and we have never had to use these procedures, and we never will." His comment was made, Leighton worked
out later, right around the time TE901 hit the mountain.

Once he learned of the missing aircraft, Constable Leighton realized with a jolt that he might be called up to assist with a massive
body recovery effort at an inhospitable, impossibly remote crash
site. Unfamiliar with snow, the youth was completely unprepared
for the possibility of abruptly leaving home and hearth in familiar

New Zealand for the nearly uninhabitable frozen wasteland of Antarctica. Soon enough, though, he was boarding a cold, uncomfortable, very noisy RNZAF C-130 Hercules headed for the ice.

Once there, Constable Leighton was given at McMurdo Station a mandatory short course in technical and survival skills, including how to pull a corpse up out of a crevasse and how to use an ice axe to stop your slide down a steep slope if you lost your balance. Afterward, he was put on the first helicopter heading from McMurdo to the crash site a forty-minute flight away. With him were Sergeant Greg Gilpin, who had been made site coordinator of the recovery effort on the mountain, and Inspector Bob Mitchell, the officer in overall charge of the Antarctic phase of what had been christened Operation Overdue. Since no landing pad had yet been constructed, the three men had to jump out while the helicopter hovered above the ground and slide down the slope.

Being the first police officers to arrive at the crash site, they undertook a preliminary inspection under the watchful eye of mountaineer John Stanton, who was responsible for their safety. At the point of impact, the three police officers were struck to see "a perfect imprint of the underbelly and wings of the ill-fated DC-10 in the snow." The aircraft clearly had been "pulling up at the time of impact and disintegrated as it travelled up the slope of the mountain." The scene was one of utter devastation on a scale almost incomprehensible to the human mind. Death had overtaken everyone on board so swiftly that no one had had an inkling that anything untoward, not to speak of fatal, was occurring. The tremendous pressures and counterpressures created when 250 tons of aircraft (and its contents) collided with the rock-hard slopes of Mount Erebus had resulted in titanic shock waves, and it was the initial shock wave that killed everyone on board.

Technically speaking, what happened was horrifying yet painless. It has been reliably calculated that, as soon as the shock waves commenced after the aircraft collided with the mountain, they were moving at a speed of around six hundred miles per hour. The first wave would have traveled through the DC-10's floors and up the legs and torsos of crew and passengers alike, where it would have been "detected by the nerve bundles." *Their* signals traveled at only ninety miles per hour, however. This meant that before the nerve bundles could communicate with the cortex, "the shock wave would have shattered the intricate network of electro-chemical interconnections." In a millisecond, the "entire biochemical structure" of everyone's brain would have been destroyed. In short, they were clinically dead not before they understood what hit them but before they even knew that they had been hit.

The camp site, such as it was at this early stage, was small and disconcertingly close to the wreckage. No sooner had Constable Leighton and Sergeant Gilpin moved into their assigned polar tents than, without warning, "a severe storm blew up with gale force winds and snow." With the windchill factor, the temperature plummeted to around minus forty degrees. Inspector Mitchell had already flown off, and the deteriorating weather now prevented others from flying in. Stranded alone on the mountainside, the police officers were aware that such storms could persist for up to two weeks. "It was no game," Leighton thought, worried about what would happen if they ran out of food. The tempest proved an omen of things to come. During the entire operation, sudden violent storms would subject the crash site to powerful swirling winds that sent jagged pieces of aircraft wreckage careening at the policemen's bodies.

Surveyors had been summoned to grid the crash site. Black flags on tall bamboo stakes marked out the grid, with each grid being roughly ninety-eight feet by ninety-eight feet. In four columns, the grid extended from the point of impact at the bottom all the way up to the last trace of wreckage at the top of the crash site. New

Zealand author Sarah Myles, whose grandfather perished in the accident, describes the grid's layout this way: "The bottom square in the left-hand corner of the crash site was numbered 1.1, then moving left to right they were 1.2, 1.3 and 1.4. Row two was then 2.1, 2.2 and so on, all the way up to the top right-hand corner, 23.4."

To process whatever was found within the gridded area, four recovery teams were created, each consisting of two police officers, a mountaineer, and a photographer. Constable Leighton was on Team Two. Their instructions were simple: "locate, tag, photograph, bag." All recovery team members had been warned that red flags in the work area indicated deep, lethal crevasses, while green flags marked body parts. They had also been advised not to handle any oxygen canisters lying in the snow because they could explode. Once on the crash site, the teams would discover for themselves that it reeked of the smell of kerosene fuel, which permeated everything, even skin and food.

"We were assigned a grid by the site coordinator, Sergeant Gilpin," Constable Leighton would later recall. "It was our job to remain in that grid and process every piece of human remains we found, no matter how long it took." Time nonetheless was of the essence because McMurdo Station's ice landing field, which seasonally could accommodate appropriately equipped aircraft, was expected to break up in about two weeks.

The job assigned the New Zealand police officers would prove punishing in the extreme. Physically, Constable Leighton found the situation was bordering on unendurable. The cold "penetrates through layers and layers of clothes" and "hurts your lungs when you breathe." The process of "locate, tag, photograph, and bag" turned out to be arduous work, from chipping bodies and body parts (each of which was counted as a "body") out of the ice to placing them in body bags and carrying those bags to just beyond the wreckage site for later helicopter retrieval.

As each body or body part at the crash site was located, a bright green temporary paper identification tag was tied with twine to it. "DEAD" was written boldly across the tag because this DVI protocol, although unnecessary on Mount Erebus, helps disaster triage teams in general to work efficiently. On the tag was then recorded several numbers that, together, comprised a particular body's identification number. The digits noted what grid a body was found in, which of the four recovery teams handled it, and the number of the body worked on in that grid.

Besides extreme physical challenges, there were psychological ones created by the ghoulish nature of the body recovery work the DVI squad was expected to perform. Constable Leighton and Sergeant Gilpin were finding it increasingly difficult to bag severed heads and contorted frozen limbs as one frightful day of doing so oozed freakishly into another. It didn't help that, morbidly, the body bags were transparent, making it impossible to forget these were only recently parts of living human beings. Constable Leighton made matters worse for himself by browsing through diaries scattered about. The last words in an entry extolling the beauty of Antarctica read, "Gee, it's great to be alive."

Particularly disagreeable was the need to remain in their filthy, stinking clothes for the duration. Human remains infiltrated their gloves, which were too short in the wrist. "We couldn't take them off because we would get frostbite," Constable Leighton explained. The team had been promised additional gloves, but they never arrived. As a result, Constable Leighton and the others were constrained to use the same pair of gloves to handle corpses and to eat their meals.

A week after TE901's collision with Mount Erebus, Constable Leighton and Sergeant Gilpin were still engaged in the macabre assignment of laboriously excavating, tagging, photographing, and bagging frozen bodies and body parts. Afterward, cargo nets dangling from support helicopters would transport the frozen flesh off the mountain. From McMurdo, the body bags would be flown to Auckland. There, at the University of Auckland Medical School's new mortuary, a second phase of Operation Overdue would commence. An entirely new suite of experts—dentists, police fingerprint experts, embalmers, and others—would then labor alongside pathologists to identify and reassemble individual passengers from body parts retrieved from up to three different body bags. It was "a gruesome human jigsaw puzzle," someone would say afterward.

In the meantime, as the work of the DVI squad continued conscientiously to proceed, two other groups possessing their own accident-related agendas came and went from the mountainous crash area. One group consisted of concerned American experts from the National Transportation Safety Board (NTSB), the Federal Aviation Administration (FAA), McDonnell Douglas Corporation (the DC-10's manufacturer), and the General Electric Corporation (maker of the engines). Their fear was that there might have been some type of technical problem with the aircraft. In eight years of flying, DC-10s had already been responsible for nearly a thousand deaths in several incidents. The initial incident, in June 1972, caused no loss of life owing to quick-thinking pilots. It occurred over Windsor, Ontario, when a DC-10 out of Detroit suffered an explosive decompression after a cargo door blew off the jet. The NTSB recommended modifications to the cargo door locking system, but they were not mandatory. Two years later, in March 1974, a DC-10 out of Paris also lost a cargo door shortly after takeoff and rapidly decompressed—but this time the crisis resulted in a crash that killed all 346 on board.

Finally, just six months before TE901 took off with Captain Jim Collins at the controls, a DC-10 out of Chicago lost its left engine and pylon assembly during the takeoff roll. The crew continued the takeoff but could not control the aircraft once aloft, leading to the fiery death of its 271 occupants and the grounding of DC-10s worldwide. Once the FAA determined that maintenance issues, not the aircraft's design, were at fault in the Chicago crash, DC-10s returned to the skies. In the short term, their reputation for safety and reliability had been considerably tarnished.

McDonnell Douglas and GE personnel must have been greatly relieved when it was determined that Captain Collins's DC-10 was functioning perfectly at the time of impact. It was now possible authoritatively to rule out technical failure as the accident's cause. A bewildered public was left to wonder what else might account for a monstrous disaster claiming the lives of 257 people in a forbiddingly distant and hostile part of the planet.

The mission of a second fast-arriving group on the mountain would be to answer that precise question. Led by New Zealand's chief inspector of air accidents, Ron Chippindale, the elite team of disaster investigators swarmed over the crash site collecting evidence that would, it was hoped, lead to a plausible explanation of what had gone so horribly wrong with the flight of a DC-10 belonging to the country's own prestigious national carrier.

6

VANISHING EVIDENCE (ANTARCTICA)

THE PTSD-INDUCING BODY RECOVERY EFFORT seemed never-ending to those involved. It was now day eight, and Constable Leighton and Sergeant Gilpin were about to make two startling discoveries. The pair was working toward the top of the crash site in the vicinity of TE901's mangled cockpit. By virtue of its having broken away from the rest of the fuselage, the cockpit had managed to land far enough afield to escape the conflagration that destroyed most of the DC-10 after impact. Unexpectedly, the pair came across a man in a short-sleeved shirt with four gold stripes on each shoulder some fifty-five yards below what remained of the flight deck. It was Captain Collins. Nearby was first officer Cassin. Compared to other bodies, the captain's, which was lying prone on the snowy slope, was "virtually unscathed."

While Sergeant Gilpin was bagging Captain Collins's body, Constable Leighton discovered near the flight deck itself a small, black ring binder with the captain's name and address on the cover. Miraculously, it was undamaged. Thinking the ring binder might contain clues as to the accident's cause, the two police officers perused its contents, which were as well preserved as the cover itself. There proved to be thirty-odd pages, of which only the initial ones contained notes. These they judged to be navigation related: radio frequencies, coordinates, and other sets of numbers. Correctly assessing it to be a critical piece of evidence in solving the mystery of what caused 257 souls to lose their lives on a godforsaken Antarctic mountain, Gilpin and Leighton carefully sealed the ring binder in a small plastic bag. After placing the parcel in a large plastic bag reserved for important pieces of evidence, they left it at the designated helicopter pickup point at the crash site for delivery first to McMurdo and thence to Auckland.

Unfortunately, there would be no chain of custody for the ring binder or any other recovered item belonging to the aircrew once it left the mountain. Police officer Gilpin would be horrified to learn *two years later* that the ring binder's contents had long since disappeared. Per their instructions as strictly body recovery workers, he and Leighton had followed the proper protocol for turning in their important find. They had expected it would be carefully examined for whatever light it could throw on the horrifying air disaster in Antarctica. While the ring binder, its contents intact, did make it back to Auckland, Gilpin had not known that afterward an Air New Zealand pilot had destroyed the ring binder's contents because of their alleged *illegibility.*

Pages from his ring binder were not the only things belonging to Captain Collins discovered at the crash site and handed in by conscientious personnel, only to mysteriously disappear somewhere between there and Auckland. Another was the captain's intact, totally undamaged flight bag with his name stamped in gold letters at the top. Mountaineer John Stanton was, like policemen Gilpin and

Leighton, among the very first to arrive at the scene of the accident. According to a sworn statement made years later by the climbing buddy with whom Stanton shared the intelligence at the time of the disaster, the mountaineer had found Captain Collins's flight bag, inside of which were a *New Zealand Atlas* and either a chart or a map. Like Gilpin and Leighton, Stanton had turned in these critical finds and later became "very concerned that the items had not been presented in evidence to the Royal Commission and had apparently not been made available to the investigators." Stanton could see why some were alleging bad faith—indeed, a cover-up— by formerly respected Air New Zealand.

Already at the crash scene, NZALPA accident investigator Peter Rhodes had developed grave misgivings about how recovered documents were being collected and accounted for. Instead of being turned over to Chief Inspector Ron Chippindale, they appeared to be winding up in the custody of his omnipresent technical advisor, Captain Gemmell. Under cross-examination later, Gemmell would initially deny but eventually concede that he had, in fact, seen Captain Collins's flight bag but insisted it was empty. The flight bag itself would make it back to Auckland and be turned in to either the police or the airline. It was last seen sitting amid other personal property at a mortuary. From there, it vanished.

If the atlas in Captain Collins's flight bag had been made available to investigators, it likely would have contained notations jotted down at his briefing nearly three weeks before the fatal flight. These notations would have proved that the airline's briefing officers' description of the final leg of the outbound trip *was* up McMurdo Sound. Since Chief Inspector Chippindale was made aware of the sticky problem of the last-minute changed coordinates only after he returned to Auckland from the crash site, it's understandable that he lost no time calling on Captain Collins's widow to ask for the atlas. Maria searched the house in vain, innocently volunteering to borrow a copy from a friend for him since obviously her husband must have taken his own on the Antarctic flight.

Long was Air New Zealand's reach, extending as it did in the person of Captain Gemmell all the way to the impact site in Antarctica. To what extent the chief inspector's technical adviser took advantage of his unsupervised presence on Mount Erebus to remove evidence—or even to plant it—is an intriguing question. NZALPA's accident investigator Rhodes was in a position to know. He would testify later at the Royal Commission hearings that he had seen Captain Gemmell with a plastic bag stuffed with documents.

Certainly, it was most odd that numerous pieces of evidence in favor of the pilots that might have turned up on the mangled DC-10's flight deck did not do so, while three documents insinuating pilot negligence allegedly did. Among the documents that would have confirmed that the pilots understood their nav track to be taking them up McMurdo Sound, per their earlier briefing, were Captain Collins's *New Zealand Atlas,* the small and large maps on which he'd been working the night before the accident flight, and an additional group of maps from his bookcase at home. There would also have been documents (including the flight plan) from the November 9 briefing and the captain's ring binder notebook with its revealing contents intact.

Even though Chief Inspector Ron Chippindale's team retrieved not a shred of documentation belonging to the dead aircrew, it did claim to have discovered three items on or around what remained of the flight deck; they all supported a theory of pilot error. The strangest of these documents was a copy of a 1977 Antarctic flight plan. In all, over their three years of operation, the Air New Zealand charters confusingly featured four different destination waypoints: Williams Field, the nondirectional beacon (NDB), the Dailey Islands, and the TACAN. The first, second, and fourth were all situated directly behind Mount Erebus. The third, however, which was the destination waypoint Captain Collins and his first officer had been briefed on, was out in McMurdo Sound off the coast of Victoria Land, in a cluster of volcanic islands.

The route ending at the Dailey Islands waypoint had been adopted by the airline in 1978. The seven Antarctic tourist flights immediately preceding that of Collins and Cassin had all carried a computerized track whose outbound terminus was there. The accident flight's pilots had been briefed on this route and justifiably expected to fly it. Why on earth would they have dug up an obsolete flight path from two years earlier to carry with them on their own journey? It was true that flight engineer Brooks had manned an earlier aerial tour, in 1977, when the official route was still over Mount Erebus. However, his flight plan's destination waypoint was the NDB, while that of the flight plan supposedly found at the crash site was Williams Field. This meant the document allegedly found at the crash site could not have belonged to Brooks.

Presumably to support the notion that pilots Collins and Cassin were aware, or should have been aware, that their nav track would take them over Mount Erebus, there was also allegedly recovered from the ice a track and distance diagram that, if plotted on a topographical map, showed that route. The problem with "finding" it at the crash site in late 1979 was that as soon as the Mount Erebus route had been supplanted with one up McMurdo Sound (over a year before the accident), the original track and distance diagram had been supplanted as well.

In hindsight, it is apparent that the interests of Chief Inspector Chippindale's team and those of the conscientious members of the DVI squad were at odds. The general aim of the government's in-house "fact-gathering" mission seemed to be badly off course.

Instead of collecting evidence, the team was evidently destroying it. The faraway crash site, which was below a subsidiary caldera of Mount Erebus, was not the only place harboring incriminating material, however. Even as Chief Inspector Chippindale and his technical adviser busied themselves at the bottom of the world, parallel efforts to uncover and destroy evidence damaging to New Zealand's national airline were underway back in Auckland.

7

VANISHING EVIDENCE (NEW ZEALAND)

WHEN INTERVIEWED ABOUT THE EREBUS DISASTER
at the end of his long life, Captain Ian Gemmell would claim that
Air New Zealand's senior officers had merely tried to control
when—not *whether*—news of the destination waypoint's changed
coordinates would be released. He himself did not consider that a
cover-up. It was a fraught time, he reminded filmmaker Charlotte
Purdy, who was the niece of TE901's deceased flight engineer
Brooks. Both he and CEO Davis were understandably bent on
protecting the airline's reputation.

They weren't the only ones. Aviation enthusiast Prime Minister
Muldoon also heartily wished to "protect" Air New Zealand—of
which he was nominally the sole shareholder—by forestalling a
finding of gross negligence against it. Besides besmirching the airline's spotless reputation, such a finding would potentially expose
the company and its underwriters to significant claims for damages

from the accident victims' relatives. It could even possibly imperil the ruling party's prospects at the polls, and Muldoon liked his job.

It would take some time for Morrie Davis and Rob Muldoon, who were of similar abrasive temperament and boon drinking buddies, to work out the details of their stonewalling strategy. Nonetheless, within hours of learning the accident's cause, the former took his first steps toward insulating the airline in a cocoon of invincibility. For now, if not indefinitely, Davis would hide its culpability from the world by keeping the undisclosed, deadly last-minute change in the destination waypoint's coordinates a secret. The airline's top officer began by keeping it a secret from the government's own air accident investigator, who flew to Antarctica the day after the tragedy to investigate in total ignorance of this critical detail. He kept it a secret from his own board of directors by telling them, seven days after the disaster, that the reason for the accident was still unknown.

He even kept it a secret from the several members of his own company deputized to assist the NTSB with the daunting job of interpreting and transcribing the cockpit voice recorder, or CVR. Both so-called black boxes had been swiftly retrieved from the wreckage. Because, however, the CVR tape would prove to be of very poor quality, it was unfortunate that the specially selected trio who left Auckland for Washington, DC, did not have the benefit of knowing TE901's flight track had been rerouted into the path of a volcano just hours before takeoff.

Several days after the accident, CEO Davis delivered a personal message via the airline's in-house journal. Published under the bestirring title "Life Must Go On," he declared it was "only right and proper" that an investigation into the accident's cause be launched. Davis cautioned, though, that irresponsible speculation on the subject was bound to occur. He then underscored his own complete confidence in Air New Zealand's "continuing operational integrity."

First officer Cassin's widow, Anne, had grown up flying with her father, John Munro, who worked in the 1950s for small airlines in the North Island. In 2002, she recalled that as a teenager she accompanied him on paper drops. "I had some draughty and turbulent flights helping my father drop the papers. On his command I would throw small loads out the doorway or push larger bundles out with my feet. My father used the bomb racks for the biggest bundles." By adulthood, Anne was married to an Air New Zealand pilot and had become a private pilot herself. Her life was deeply entwined with the aviation industry.

Only three hundred feet from the St. Heliers beach, the Cassin home offered sensational harbor views and marvelous boating opportunities for those who, like first officer Cassin and his spouse, possessed watercraft. On November 29, 1979, the day after the accident, throngs of people began descending on the petite home to offer the young widow condolences and support. Several were close relatives. The airline flew first officer Cassin's parents and brother up from Napier, for instance. It also flew Anne Cassin's sister and brother-in-law up from Christchurch.

Scores of visitors from the airline and its regulator swamped the tiny house as well. Captain Peter Grundy, now flight operations manager, was one of the first to pay a condolence call. He had been involved in the early planning of the Antarctic flights with Captain Gemmell. Another was the president of NZALPA, who introduced Anne Cassin to Captain Bruce Crosbie, the member who had been designated to liaise with the dead crew's families on that organization's behalf.

Because many callers at the Cassin home wanted to know where McMurdo was, Anne casually started directing them to the briefing notes and maps that were in a folder sitting on the coffee table in her small living room. These documents were in the family residence, not at the crash site, because first officer Cassin had left them at home by mistake the day before—that is, the morning of the accident flight. As a private pilot herself, Anne had studied the folder's contents when her husband returned home from the briefing session nearly three weeks earlier. At the time, she had not been especially interested in the flight documents. Anne did notice, however, that the folder included three additional pages of personal notes Greg had jotted down concerning the upcoming trip.

Anne Cassin and Captain Bruce Crosbie did not hit it off. According to Justice Mahon's recollections from an interview with her after his official report came out, she was "resentful at the nature of the inquiries" Crosbie had made. For some reason, the liason officer seemed "intent on discovering what information she might have in her possession with regard to the fatal flight." Anne also soon discovered that "everything she told him was very shortly within the knowledge of the airline management." Why would *airline management* be receiving intelligence reports rather than NZALPA, whose representative he officially was?

In her shocking testimony at the Royal Commission hearings on November 21, 1980, Anne testified that it was not she but her brother-in-law who had turned over to Captain Crosbie all her spouse's flight papers, as well as private documents and even clothing. "I was not in the house at the time. Taken without my knowledge or my permission. I did not find out until a few days later." Anne said that the liaison officer "denied having seen the briefing notes." However, when he was in the Cassin home, she remembered "him looking at them." She added that Captain Crosbie came to her home "every day at that stage, and [the folder containing the briefing notes] was always there."

In his own testimony before the Royal Commission on December 3, 1980, Captain Crosbie stated that Anne Cassin's brother-in-law had requested that he "remove F/O Cassin's fly-ing manuals and articles of uniform." These had been "placed into cardboard boxes" for him. The liaison officer admitted that Anne "wasn't aware that I was receiving the manuals and uniform."

Besides being angry that her brother-in-law would usurp her authority as homeowner for the purpose of dealing covertly with the NZALPA liaison officer, Anne was confused. Without under-standing why, she now realized that she should have more carefully guarded her husband's flight-related paperwork. After her visit-ing relatives had returned home, Anne thought to check on her husband's briefing session documents specifically. The folder was still lying out on the coffee table, and several innocuous papers remained inside, as if to provide heft. However, the folder's vital contents themselves had vanished. Why? At that stage, it did not occur to her—or to Maria Collins—that the airline was planning to argue that the pilots *knew* their route went over Mount Erebus, for which reason all physical evidence to the contrary naturally needed to be destroyed.

Someone seemed to be going to a lot of trouble to make sure no evidence that might incriminate Air New Zealand would ever turn up at the pilots' residences. Although Anne Cassin's home had already been sanitized, it was nonetheless covertly broken into later and searched. So was the Collins home. Oddly, the break-ins hap-pened on the single occasion that each widow left her house over an extended period. Some speculated that both properties must have been continually monitored by an organization with the where-withal to do so and the flexibility to strike fast when presented with an opportunity to get in and out undetected. Especially curious was the illegal entry into the Collins dwelling, in which power to the house was cut and a file of accident-related papers, maps that Jim had collected over the years, and correspondence were taken. A photo of Captain Collins had been torn to shreds as well.

The cases were never solved. However, Prime Minister Muldoon's penchant for counterpunching enemies past, present, and future was well known. So was the fact that he oversaw New Zealand's Security Intelligence Service (SIS). Having both the will and the means, the high-handed Muldoon was thought by many to have ordered that agency to conduct the break-ins.

———————————

An essential aspect of Morrie Davis's effort to hide the coordinate change was the collection and impoundment of Antarctic-related documents, particularly those relating to the fatal flight, wherever they might be found. Although some were undeniably on Mount Erebus in and around the crash site, plenty were stored in Air New Zealand's own headquarters on the Auckland waterfront. This uncomfortable fact inspired the carrier's CEO to launch a company-wide campaign to destroy all but a sole copy of the airline's paperwork relating to its polar sightseeing flights. "I required one single file authenticated and unadulterated," Davis explained. Surplus copies along with documents not deemed pertinent were to be eliminated forthwith. His confiscatory policy resulted in all kinds of potentially revelatory documents being turned in and never seen again, including Captain Ross Johnson's Antarctic briefing file.

The campaign of document destruction was run by the airline's top safety officer, George Oldfield, who regarded shredding as "the most practical" means of getting rid of records. He got to work eliminating every extra, spare, or duplicate copy of all the Antarctic flight documentation and other airline material as well. Maurice Williamson, who worked at Air New Zealand during the period 1975–1987, would recollect years later that entire boxes of papers were dumped in the shredder. "Why were people bringing in boxes and why were people down[stairs] on Flight Operations running around getting files?" the bemused corporate planner wondered at the time.

Perhaps Williamson can be forgiven for initially assuming the frenzied activity was designed to make material available to probing civil aviation authorities rather than available for shredding. Maria Collins would discover from a friend who worked on the same floor as the office in which the documents were being destroyed that so relentless was the attempt to eliminate records pertaining to the chartered flights that someone's tie got caught in the shredding machine. The upshot of this unprecedented undertaking was that all kinds of airline operational files that Justice Mahon would normally expect to review during an investigation would prove unavailable once the Royal Commission commenced. "The relevant parts of the Commercial Affairs Division files, the Flight Operations Antarctic file, the Navigation Section files dealing with Antarctic flights and so on," he would find, were simply gone.

Incidentally, during the Royal Commission hearings, Justice Mahon would become suspicious that every document he *did* receive from Air New Zealand was only a photocopy. Because the detection of small changes is easier on an original document than on a photocopy of it, he wondered how many tiny alterations might have been made on the photocopies he'd be given in lieu of original documents. "This was at the time the fourth worst disaster in aviation history," he declared, "and it follows that this direction on the part of the chief executive for the destruction of 'irrelevant documents' was one of the most remarkable executive decisions ever to have been made in the corporate affairs of a large New Zealand company."

At the same time as valuable clues to the cause of TE901's collision with Mount Erebus were methodically being either destroyed or reduced to photocopies, Davis issued a prohibition against airline employees communicating with the press. On the one hand, this injunction may have arisen in part from an understandable company impulse not to allow the line pilots to jeopardize Air New Zealand's position with respect to its passenger liability insurance, which could be limited or unlimited depending on what was established to be the predominant cause of the disaster. (Liability would be limited if the crash was attributed to pilot error, unlimited if deemed to have arisen through airline negligence.) On the other hand, Davis very conspicuously possessed a combative, dictatorial type of personality, and one of his "continuing preoccupations," according to Justice Mahon and others, was "the veiling of [the airline's] commercial operations from public scrutiny." In the wake of TE901's demise, there was much to veil.

In keeping with his "hang tough" approach to requests for information regarding possible reasons for the crash, CEO Davis was careful in interviews never to mention the matter of the aircraft's altered coordinates for the flight's destination waypoint. However, in February 1980, an Auckland paper alleged both that the aircraft's computer track had been altered just prior to the fatal flight and that the crew had not been informed of it. In a statement flatly denying any incorrect data had been fed into the DC-10's navigation computer, Davis declared, "The navigation information and flight plan for the aircraft which crashed was accurate and entirely in order." He could say this with a straight face because of his idiosyncratic use of the word "accurate." Davis's position was that the navigation track had indeed been changed—but changed from an *inaccurate* route back to an *accurate* one.

The deception involved here was to label as inaccurate Air New Zealand's Antarctic flight path in use for the previous fourteen months: the path up McMurdo Sound. The CEO's premise was that the route on record for *1977*—the direct route over Ross Island's formidable Mount Erebus—was the accurate one. In Davis's specious reasoning, it followed that a relocation of that so-called

inaccurate route up the sound back to one over Erebus "was accurate and entirely in order." Being an authoritative statement by
the airline's CEO, this deceptive language was taken to be the last
word on the matter by many people, quelling public suspicions for
the time being.

PART TWO

DUELING ACCIDENT REPORTS

8

WHEN ENDS
JUSTIFY MEANS

IN THE CASE OF AVIATION ACCIDENTS, WHICH HIS-
torically had involved small aircraft, it is in New Zealand the job
of the government's Office of Air Accidents Investigation to determine what happened. The crash involving Captain Jim Collins's
planeload of happy sightseers was not a run-of-the-mill accident,
however, but the world's fourth-worst aviation disaster up to that
point. It was "singularly unfortunate," in Justice Mahon's retro-
spective view of the matter, that while the Office of Air Accidents
Investigation was statutorily independent of the Ministry of
Transport's Civil Aviation Division, it was in fact part of it. In other
words, given that Civil Aviation was the airline's regulator, the chief
inspector of air accidents, Ron Chippindale, had a built-in conflict
of interest before he took a single step toward launching his inquiry.

Ron Chippindale had served thirty years as a pilot and officer
in the Royal New Zealand Air Force (RNZAF) before joining

the Ministry of Transport as an air accident inspector in the mid-1970s. When the current chief inspector retired not long afterward, Chippindale had taken over as his successor. About the Erebus crash, which involved a jetliner rather than his usual fare of small planes, he observed, "It has been hard to establish a definite cause. With a structural failure in the aircraft, it would be quite simple." Despite the accident's apparent complexity, Air New Zealand and the government that owned it anticipated findings from Chief Inspector Chippindale that would account for the horrific disaster in Antarctica in a way that would not reflect badly on the airline.

Meanwhile, mounting pressure exerted by an aroused public had constrained the highly resistant Prime Minister Muldoon to agree to a second investigation, one that would be perceived as independent of governmental interests. In selecting the High Court's distinguished Justice Mahon as royal commissioner to conduct a one-man second inquiry, the prime minister had followed the advice of his trusted personal attorney, Des Dalgety, as well as the advice of distinguished Auckland attorney Lloyd Brown, who had already been chosen to represent the airline as lead counsel in the public inquiry. It was thus assumed at the highest levels that, as the head of a Royal Commission of Inquiry into the colossal Antarctic air disaster, Justice Mahon would view matters in the proper light.

Since such an appointment was by special warrant, Attorney General McLay was the official who, on April 21, 1980, at Prime Minister Muldoon's behest, entrusted Justice Mahon with the job of launching a second official inquiry into the grisly deaths of 257 people at the desolate bottom of the world. In a press statement, Muldoon explained his preference for a single commissioner over a committee. He further declared how challenging it had been to find a man of integrity who could master the technical details of a catastrophic air disaster yet who had no previous association with Air New Zealand. The prime minister seemed well satisfied with his selection, and there can be little doubt that Justice Mahon too was initially pleased with his novel assignment.

It was not accidental that the newly designated royal commissioner was not empowered to begin hearing evidence until nearly three months after his appointment. According to Attorney General McLay, the delay was to allow the government's own air accident specialist time to complete *his* internal investigation and publish a report revealing *his* findings. Postponing the Royal Commission's start date would give those downplaying the relevance to the crash of a last-minute change to TE901's flight path, undisclosed to the aircrew (and later the public at large), a small window in which to finalize and disseminate as widely as possible, in what would become known as the Chippindale Report, the government's own views on the cause of and culpability for the horrific air disaster.

That the Muldoon administration could even contemplate proceeding in this potentially divisive fashion was owing to a quirk in the country's legal system. Unlike in England, where a call for an independent inquiry generally forestalled publication of any governmental findings, in New Zealand it did not. The prime minister and his top associates' contemplated action raised eyebrows in the legal community and triggered considerable public opposition. But rushing the Chippendale Report into print to broadcast its findings before the second inquiry could even begin—much less conclude in a report that might reach different conclusions—proved irresistible to Prime Minister Muldoon. He was known to enjoy what he termed "counterpunching" active and potential opponents alike.

While Muldoon undertook measures designed to blunt potential criticism of his beloved airline, that airline was launching a publicity campaign of its own. Journalist and communications expert Ken Hickson was an experienced aviation writer working in-house for Air New Zealand at the time of the disaster. His book *Flight 901 to Erebus*, published later in the same year as the official report of the government's chief inspector of air accidents (1980), hosannaed that report's author. While Hickson's book did provide informative background detail on Air New Zealand's Antarctic flights, its real objective was to extol the unparalleled investigatory skills of Chief Inspector Ron Chippindale. When, for example, Attorney General McLay announced that there would be an independent Royal Commission of Inquiry into TE901's crash, Hickson told his readers that he immediately went on television's *Eyewitness* program to testify that he "didn't think an [additional] inquiry into the accident was necessary." In Chief Inspector Chippindale's "methods and expertise," Hickson "had complete faith."

So complete was that faith that on the very day the Chippindale Report was released to the public, *Flight 901 to Erebus*'s author participated in a *Radio New Zealand* roundtable devoted to a discussion of the government's findings. During the radio show, he again underscored the chief inspector's "competence and authority," reiterating that his report rendered a second investigation superfluous. The chief inspector had already thoroughly covered all facets of the accident and made a sound determination as to its cause.

When Ron Chippindale was informed that TE901 was probably lost, he had immediately assembled a team of investigators to assist him in probing the suspected disaster. Designating Captain Gemmell, who was the real architect of the Antarctic flights, his technical advisor, he and Gemmell traveled together to Antarctica the next day, November 29, 1979. At the crash site, a potential crime scene, Captain Gemmell was able to move about with complete freedom. Later, when the so-called black boxes were returned to New Zealand after official analysis at NTSB headquarters in Washington, DC, Chief Inspector Chippindale would share their contents with Captain Gemmell at his private residence in Wellington. The stakes for the airline generally, as well as Captain Gemmell personally, were very high, and the latter's relationship with the man running the government's investigation was obviously cozy. There would have been numerous opportunities for Captain Gemmell to influence Chief Inspector Chippindale's views on accident causation. As time would show, the latter proved an eager pupil.

Neither before nor during his investigations at the crash site, apparently, was Chippindale told of the last-minute coordinate change that was made without notifying TE901's pilots. Once returned from Antarctica and informed of the airline's misstep, he was prepared to dismiss it as inconsequential. He'd already arrived at his own theory of causation, which he lost no time sharing with Captain Collins's widow two weeks after the accident. "Jim was too low, Maria," Chief Inspector Chippindale had intoned. "Jim was too low."

It was Chippindale's position that the aircrew had behaved negligently in multiple regards, each of which contributed to precipitating the accident. There was a single supervening cause, however: Captain Collins's decision at the end of the outbound flight to descend below 16,000 feet. This number was the minimum safe altitude for Air New Zealand's Antarctic flights that had been approved by its regulator originally (1977), and it was still formally on the books. That specific altitude had been established in conjunction with a flight path that Civil Aviation had concurrently approved. It too was still formally on the books. That original path directed an airliner into Lewis Bay and over Ross Island. It was to avoid colliding with the 12,450-foot active volcano on that track that the 16,000-foot minimum safe altitude regulation had been created.

There is a certain superficial logic to the argument that had Captain Collins complied with the old 16,000-foot rule, he would not have crashed into Mount Erebus. Why would he think to do that, though? For the previous fourteen months, all Antarctic charters had been programmed to proceed at the end of their outbound journey up the flat sea ice of McMurdo Sound to the Dailey Islands. From their preflight briefing, Captain Collins and first officer Cassin expected to follow that route as well. Moreover, pilots headed for Antarctica in the second half of 1977, and all of 1978 and 1979 too, had been allowed to fly as low over the sound as the local Americans' air traffic control officers deemed safe (typically, 3,000–1,000 feet).

Chippindale's definition of "probable cause" omitted all antecedent considerations to focus exclusively on "the last thing which occurred in the sequence of the flight which made the accident inevitable." By that standard it was obvious that the 250-ton DC-10 crashed because it was "too low" to avoid smashing into the mountain. The real question, however, was why this airliner was in Lewis Bay at all, headed directly at an active volcano on Ross Island, when for the past fourteen months all Antarctic sightseeing tours had been programmed safely to fly up the broad expanse of McMurdo Sound well to the west of the mountain.

One of the chief inspector's principal findings was that the undisclosed last-minute change in TE901's flight plan—which sent the aircraft not up McMurdo Sound but over Ross Island on a beeline for Mount Erebus—did not mislead the pilots. One might be forgiven for imagining that the fatal accident itself constitutes proof that they were misled. According to Chippindale's reconstruction of events, however, Captain Collins and first officer Cassin had been informed at their preflight briefing that their computerized Antarctic track would be over Mount Erebus—that it *always* went over Mount Erebus. There was absolutely no evidence, the chief inspector insisted, that the pilots did not know they were flying into Lewis Bay on a track leading toward the mountain. All they needed to do was responsibly fly over it rather than into it.

In making these incendiary charges, Chippindale had let himself get a bit carried away. For one thing, there were three Air New Zealand pilots still alive who had attended the same briefing as Collins and Cassin and knew its contents were not as the chief inspector described. Might they not speak out publicly at some point? For another thing, there were three unimpeachable witnesses to the fact that TE901's captain had spent the evening prior to the fatal flight meticulously plotting his route up McMurdo Sound on topographical maps he'd acquired for just this purpose. These witnesses were Jim Collins's wife and the two eldest of his four daughters.

"He told me the route he was going to take, and it didn't add up to where the plane ended up," the eldest Collins daughter, Kathryn, would later recall. "It wasn't where he said he was going." She remembered her father mentioning "the Dry Valleys, the coast of Victoria Land, the Williams Field ice runway, the McMurdo Sound approach. He didn't mention Mount Erebus." The second eldest daughter, Elizabeth, remembered her father, on that last evening of his life, unfurling an oversized map and pointing at the mountainous coast of Victoria Land. He planned to keep "fairly close to this bumpy lot," she reported him as saying. According to the girls, their father was working with both a smaller map and a very large one, in addition to his atlas. The color scheme on the big map was peculiar, Elizabeth thought at the time—green for the sea, purple for the mountains. All told, the two teenagers spent about twenty minutes learning about their father's upcoming aerial tour.

Kathryn's and Elizabeth's testimony, which would eventually assume the form of sworn affidavits, was potentially explosive. If Captain Collins had indeed been plotting a route up McMurdo Sound the night before the fatal flight, he could only have done so based on latitudinal and longitudinal coordinates contained on the digitized flight plan presented to the five airmen who had attended the Antarctic briefing of November 9. During the Royal

Commission's inquiry into the cause of TE901's crash, the identity of the two maps would publicly be sorted out, as would how and when Captain Collins acquired them. The oversized one, GNC21N, proved to be an American product, about five feet long—precisely the unwieldy sort of thing a pilot would need to spread out on a dining room floor to show his young daughters exactly where his travels would be taking him the next day. The sea and ice were green and the land purplish orange on this exceedingly big map or chart.

That Captain Collins had this all-important, outsized document with him on the fatal flight is suggested by the CVR, on which there is the sound of paper being repeatedly rustled. The need to open, fold, open, and refold the mega-map as the trip progressed is the only reasonable explanation for it. Conclusive evidence exists too, thanks to Chief Inspector Chippindale. In his investigations, he had traced a radio call from one of the accident aircrew's members to a Peter Tait in Nelson informing him that the DC-10 had on board the GNC21N mega-map that he'd lent him. This extra-large map or chart is doubtless what the mountaineer John Stanton found after the crash in the dead captain's flight bag, which, like its contents, soon vanished without a trace somewhere between Ross Island and Auckland.

When Chief Inspector Chippindale returned from Antarctica, Maria Collins would later testify at the Royal Commission hearings, she informed him that her husband was plotting his route on maps the night before his flight. Asked at that inquiry why he had not included this intelligence in his official report, the chief inspector would declare that it did not constitute "evidence" as he understood that term. His notion of evidence involved only *tangible* things, and nothing in this category existed to the best of his knowledge. As for GNC21N specifically, it would turn out Chippendale did know all along that Collins possessed this huge map. He chose not to disclose that in either his report or his subsequent testimony, however, since alleging Captain Collins had no legitimate checking reference with him on the flight deck served to bolster his thesis that the pilots had violated sacred aviation industry protocol by descending to low altitude without having first fixed their position via such a technical aid.

In the estimation of colleagues who knew Captain Collins well, the only way this conscientious pilot would have remained locked onto the nav track on approach to the destination waypoint was if he had preplotted his flight path, kept monitoring his position via at least one of the maps his family had seen him working on the night before the disaster, and prior to descent positively identified where the aircraft was via geographical features in the area he was approaching. Besides, locked on the computerized route, it was impossible for Captain Collins *not* to know where he was. All he had to do was compare the input on his instrument panel's distance-to-run indicator with his topographical map. That then informed him how many miles away from the destination waypoint the aircraft was on his plotted track.

Consider this basic question: What pilot in his right mind, knowing his computerized flight path is directing his aircraft straight at a 12,450-foot mountain, would elect to descend to 1,500 feet and immediately plow into it? When Justice Mahon later put this very question to the head of Civil Aviation, Captain Kippenberger could only hypothesize that Captain Collins must have been suddenly afflicted by a mystery illness—and that first officer Cassin might have fallen prey to it too at the same time. In fact, since there were five men on TE901's flight deck at the time of impact, all would have needed to succumb simultaneously to the mystery illness for the theory to hold up.

It should be obvious that Captain Collins would only have locked back onto his nav track after his second descending orbit at the approach to Lewis Bay if he believed the infallible inertial navigation equipment had been programmed—per his preflight briefing—to proceed up the flat sea ice of McMurdo Sound. It should be obvious, but even today, over forty years since the DC-10's collision with the mountain, there are still those under the misapprehension (deriving ultimately from the Chippindale Report) that the pilots were flying in cloud, "not certain of their position." From the hundreds and hundreds of passenger photographs taken on November 28, 1979—TE901 was a *sightseeing* flight—we know there was no cloud through which they were flying. The plane was in totally clear air. And the pilots *were* certain of their position. They may have been wrong, but they were confident of being over McMurdo Sound.

Incidentally, an odd feature of the treasure trove of photographs produced at the Royal Commission hearings would be that they showed views in all directions except south. Because of Captain Collins's two leisurely descending orbits on approach to Lewis Bay, the DC-10 would have been side-on to Ross Island on four occasions. Justice Mahon thought about forty pictures toward the south should have been taken during those orbiting sequences. A leading American Antarctic expert named Charles Neider was of the same opinion. It occurred to Mahon that assuming such evidence at one time existed, it clearly could not have depicted a cloud base descending almost to the bottom of Mount Erebus's icy slopes, as the chief inspector alleged. If it had, he would have seized on that evidence instantly as vindication of his strongly held theory that the pilots were flying in or toward cloud. If, on the other hand, the south-facing shots revealed something else entirely, perhaps it would have been deemed prudent to dispose of them. In contrast, opponents of the judge's interpretation would allege that, it being "cloudy," there were no pictures worth taking.

The gold standard for transcribing a CVR is the National Transportation Safety Board in Washington, DC. Once this so-called black box and its mate, the digital flight data recorder (DFDR), were recovered and delivered to Auckland, an elite team was quickly assembled and sent to the US with the precious data. Its mission was to decipher, with the assistance of NTSB personnel and their special listening filters, what was said on the flight deck during the final thirty minutes of the aircraft's life.

Fleet manager (DC-10) Barney Wyatt, chief flight engineer (DC-10) Don Olliff, and DC-10 captain Arthur Cooper had been chosen for the transcription team precisely because of their familiarity with the crew's voices as well as for their expertise with DC-10s. Because the quality of the recording proved abysmal, the men could not initially understand much. Still, as one team member would later recall, they were shattered upon hearing their deceased colleagues' voices on the CVR tape for the first time. These were their last words.

At the NTSB, a fastidious protocol was implemented requiring all team members unanimously to agree on a word before it could be incorporated into the official transcript. A second requirement was consensus on the identity of a speaker before particular words could be attributed to one or another of the five men on TE901's flight deck. At the end of their intense, week-long effort, Cooper declared everyone well satisfied with the accuracy of the group's recording. Upon their return to Auckland, team members were given to understand, theirs was to become the CVR transcript of record. That did not happen.

It did not happen because, in a departure from internationally accepted practice, Chief Inspector Ron Chippindale now took sole custody of both the CVR tape and the Washington Transcript. We know he took them first to his home in Wellington, where, in another contravention of international protocols, he and Air New

Zealand's Captain Gemmell listened to the tape together. Whether, huddled with Chief Inspector Chippindale in his private residence, the actual architect of the Antarctic charter flights took advantage of his position as technical adviser into the accident's cause further to mold Chippindale's thinking is an engrossing question. It's entirely possible that his next highly controversial move with the CVR tape was hatched right there in his own living room in consultation with Captain Gemmell.

That move—which again defied accepted practices regarding CVR transcriptions—was for Chief Inspector Chippindale to fly with the tape to the United Kingdom. There, he listened to it *by himself* with the help of sound equipment available at RAF Farnborough, which he claimed was superior to that of the NTSB. After doing so, Chippindale produced an alternative, unauthenticated transcript of the last thirty minutes of the CVR.

It was a most unusual document, this Farnborough Transcript, containing *fifty-five deviations* from the Washington Transcript. Because Chief Inspector Chippindale attached *his* version of the CVR's content to his own interim and final reports, the government was able to bury the Washington Transcript, publish its in-house official investigator's altered version in its stead, and represent it as an authentic record. This allowed the Farnborough Transcript to gain a great deal of traction with the public.

That was regrettable because Chief Inspector Chippindale's insertions, deletions, and alterations of words being spoken on the flight deck as the aircraft approached its gruesome fate affects how one understands the aircrew's state of mind at the end. The Farnborough Transcript is at pains to insinuate that the pilots were flying irresponsibly at low altitude in or approaching cloud, with no idea where they were, while the engineers unavailingly importuned them with "mounting alarm" to come to their senses. Multiple remarks in the Washington Transcript, by contrast, confirm that the pilots were confident of their location and the engineers comfortable with the pilots' handling of the DC-10.

Aircrews in 1979, at least those of Air New Zealand, no longer operated hierarchically, with a captain's judgment ipso facto considered superior to that of the rest of the individuals on the flight deck. Instead, they now operated by means of a *crew loop*, according to which everyone on the flight deck participated in communal exchanges about a flight's progress and was responsible for speaking up if perturbed by something. Flight engineer Brooks was an instructor in the crew loop procedure himself. It's a technique that greatly lessens the chances for air accidents.

Chief Inspector Chippindale was personally unacquainted with anyone on TE901's flight deck, but Air New Zealand's Captain Arthur Cooper of the Washington Transcript team did know both engineers well. He considered it inconceivable that flight engineer Brooks would not have spoken up if alarmed. In fact, twenty-six seconds before impact, Brooks did speak up—"I don't like this"—at which point, feeding into the crew loop, Captain Collins expressed his own concern with the flight's progress and decided to fly away. The aircrew's universal approval of Captain Collins's operational decisions as revealed in the Washington Transcript is diametrically opposed to Chief Inspector Chippindale's portrait in his quirky text of an irresponsible captain hellbent on ignoring the safety cautions of his colleagues—with tragic consequences.

In a later explication of his "ends justify means" methodology in creating the Farnborough Transcript, Chief Inspector Chippindale would assert that "the purpose of the transcript" is to "support the conclusions of the report itself." Consequently, he said, "I continued editing the transcript almost continuously until the time I presented it to the government printer." One thing he "edited" was a remark made by commentator Mulgrew three minutes prior to impact: "Taylor on the right now." Victoria Land's

Taylor Valley would have been to the right of the aircraft had it, in fact, been entering McMurdo Sound rather than Lewis Bay. Chief Inspector Chippindale chose to obfuscate where Mulgrew believed the DC-10 was by turning his simple declaration into a question and fabricating Captain Collins's answer to it. "The Taylor or the Wright now or do ya?" "No, I prefer here first!"

In another instance, Chief Inspector Chippindale claimed to have made out certain words not intelligible enough to make it into the Washington Transcript—"Where's Wilson?" "Over to the right." Later, though, he deleted them from his polished transcript. Probably this was because these words provided further confirmation that the pilots believed they were in McMurdo Sound, with the highly conspicuous Wilson Piedmont Glacier off to the right stretching down the Antarctic coastline. When asked why he removed his own reference to the glacier in the final version of his transcript, Chief Inspector Chippindale responded, "We were unable to identify any place with the single name of Wilson." He would have it that the pilots were lost in cloud and speeding, at low altitude, toward a mountain on Ross Island they knew was there. His transcript of the CVR could thus not be allowed to reveal that they understood themselves to be in the sound, with clear views of "Taylor" and "Wilson."

Chief Inspector Chippindale's thesis that Captain Collins was taking his aircraft down into bad weather, where no one could see, is belied by the pilots' multiple affirmations during the last thirty minutes of the outbound journey that they were flying visually in clear air. "Bit thick here, eh Bert?" Chief Inspector Chippindale, in his transcript, has someone on the flight deck asking in a remark heard by no one on the New Zealand team working with the NTSB. Presumably the "thick" was meant to suggest conditions of poor visibility, which we know from the vast trove of tourist photographs was not true: visibility was excellent. It didn't help Chippindale's credibility that there was no one on the flight deck—or indeed on the flight—with the first name Bert. Perhaps "Bird," as in Cape Bird, which the aircrew imagined they were passing on their left as they entered McMurdo Sound, rather than "Bert," was what was said.

In short, the Farnborough Transcript represents the work of a single, conflicted individual straining to twist the facts to square with his a priori thesis of pilot error. Admittedly, the chief inspector of air accidents was in an unenviable position. While Chippindale's office was a governmental one, he was nonetheless charged with investigating possible negligence by two other governmental entities: the national airline itself and the division of the Ministry of Transport that served as its regulator. Should either be found negligent, there would be vast financial and political ramifications for the government, which presumably Prime Minister Muldoon or one of his inner circle early on took Chippindale aside to explain. Under these circumstances, the chief inspector doubtless appreciated that pilot error was the only proper conclusion for him to reach. It did not matter how much evidence had to be suppressed, contorted, or manufactured in the process: Chippindale was pleased to oblige.

A particularly misleading ancillary charge that Chief Inspector Chippindale leveled against the accident aircraft's allegedly incompetent pilots was their failure to detect the volcano via radar. Justice Mahon would later follow up on this allegation by interviewing experts at the Fort Lauderdale premises of the Bendix Corporation, manufacturer of the equipment. There he would learn the weather radar installation on a DC-10 could not have picked up the mountain. It was designed to detect moisture in clouds (to alert pilots to turbulence). There is no moisture in Antarctica, though: it is drier than the Sahara Desert. The Bendix personnel informed Justice Mahon that given the thick layer of dry snow and dry ice that coats Mount Erebus's slopes, any return would have been nil. This, they explained, was because of "the inability of the radar pulses from the radar to achieve any return once they penetrated the crystalline structure of dry snow and dry ice." The Bendix manual itself specifically advised against using its radar equipment for terrain avoidance.

Before Mahon left Bendix's premises that day, he asked whether anyone else had inquired about radar returns that might have been obtained by the accident aircraft as it approached Ross Island. Someone from McDonnell Douglas had, in fact, contacted the company, he was told. Bendix had given the DC-10's manufacturer the identical information it had just shared with Justice Mahon. How odd. In the Chippindale Report, the chief inspector claimed it was an expert at McDonnell Douglas that had informed him the aircraft's radar *would* have detected the mountainous terrain ahead. When challenged on this point later, at the Royal Commission hearings, he could not remember that expert's name.

A distinguishing feature of the chief inspector's investigative methodology was his mode of collecting and keeping track of evidence. His predilection was to seek it out privately, put nothing in writing, and keep no record of the identities of even those providing the most important intelligence upon which his interim and final reports' findings would be based. Stuart Macfarlane, the noted Erebus researcher, famously described the strength of the Royal Commission, upon whose findings the Mahon Report is based, as deriving from "the facts that it was conducted in public, the evidence was subject to cross-examination, the evidence was taken down in writing." For researchers there consequently exists in association with the Mahon Report a cornucopia of exhibits, briefs of evidence, oral examination transcripts (running to 3,083 pages), and written analyses by counsel available for independent scrutiny. The chief inspector's accident investigation proceeded along different lines entirely. His approach emphasized chatting "informally and off the record" with witnesses to extract "the maximum amount of information" bearing on the disaster's cause.

Speaking of the disaster's cause, it was strange the chief inspector did not chat informally and off the record with anyone in the airline's Flight Operations Division, especially its Navigation Section, which had committed unthinkable programming blunders. Instead, he consulted "some eighteen highly qualified international experts and aircraft accident investigators" before writing what the world knows as the Chippindale Report, which attributes the crash to pilot error. Who were these august authorities? The first named were members of Civil Aviation, the airline's regulator, which was an interested party owing to the very real possibility of culpability on its part. Also named was Air New Zealand, a second vitally interested party for the same reason. Chippindale adduced NZALPA as a third group supporting his findings, which was bizarre, given that this group would strenuously dispute them later at the Royal Commission hearings.

The chief inspector's further claim that the NTSB (and also the FAA) concurred with his views is highly suspect. It hardly seems likely the Americans would have embraced Chippindale's findings, based as they were on controversial changes the chief inspector unilaterally had introduced into the NTSB's painstaking transcription of the DC-10's CVR. Also in agreement with his conclusions, Chippindale suggested, were McDonnell Douglas, maker of the aircraft, and General Electric, which built its engines. Since mechanical failure had quickly been ruled out as the accident's cause, these companies' views, if any, on what did precipitate it were clearly irrelevant.

In conducting his research and reaching a determination as to culpability for the accident, Chief Inspector Chippindale stressed that he had not been under pressure from the Minister of Transport, the director of Civil Aviation, Air New Zealand, "or any other person or organization." His conclusions were "based on the assessment made by all the investigators involved in preparing our report." Since, though, as we've just seen, the only investigators involved (including him) were hugely interested parties, it's hardly surprising that the Chippindale Report finds no fault with the airline or its regulator—it finds fault only with the deceased aircrew.

9

QUESTING FOR TRUTH

CAPTAIN GORDON VETTE HAD FIRST LEARNED OF the disappearance of TE901 as he was preparing to fly a DC-10 from Honolulu to Los Angeles. "I was stunned, almost beyond belief," he would recall in his insightful study, *Impact Erebus.* "I had flown with every crew member over the entire period each had been employed by Air New Zealand and had watched the progress of Jim Collins for nearly twenty-five years, from the time of his induction as a student pilot in the [Royal New Zealand Air Force] at Wigram when I was one of his instructors. Some of the crew I had flown with only days earlier." Upon hearing rumors that the government's chief inspector was prepared to ignore the crew's untarnished record and summarily chalk the crash up to pilot error, he decided then and there to undertake his own private accident investigation. It would take a full year and a half to complete, "carried out between flying assignments and hedged with frustrations."

Chief Inspector Chippindale may have thought it unremarkable for a highly experienced pilot whose peers considered him the epitome of a non–risk taker to depart from his normal conservative practices and slam his aircraft into a mountain he knew was there. In Captain Vette's view, however, this would have entailed an "extraordinary degree of incompetence" on both Captain Collins's part and on the part of the rest of the crew. In contrast to Chief Inspector Chippindale, Captain Vette accepted as "axiomatic" that those on the flight deck were exhibiting their customary high standard of professionalism and conscientiousness and began searching for circumstances that could have deceived "even an experienced crew." One obvious mystery was why no one on the flight deck—four crew members as well as veteran Antarctic expert Peter Mulgrew—*saw* the looming volcano they were rapidly approaching. Something very strange indeed must have occurred to fool airmen of the quality of those responsible for TE901. Captain Vette, who, like Justice Mahon, had a strong sense of morality and fair play, wanted badly to know what it was.

Captain Vette's character had most recently been on display less than a year before the Erebus catastrophe in an odd aviation incident that occurred over the Pacific Ocean. En route from Fiji to Auckland with a planeload of passengers, the DC-10 captain had picked up a distress call from the pilot of a Cessna that was lost due to a malfunction in its automatic direction finder. Knowing the pilot would perish if he didn't receive assistance, Captain Vette diverted his large airliner to search for the tiny one, even enlisting all the passengers to help by looking out the windows. Through a series of logical navigational steps, he eventually ascertained its location, enabling the Cessna to reach its destination, Norfolk Island, just before running out of fuel. The small aircraft had been in the air for twenty-three hours. Captain Vette's heroics won the crew, which included Gordon Brooks as flight engineer, a certificate of commendation from McDonnell Douglas for "the highest standards of compassion, judgment, and airmanship." The story was later recounted in a book and dramatized in an American TV movie.

Captain Vette's sons, both psychology students, introduced him to works on visual perceptual psychology. As his understanding in this specialized area progressed and he consulted experts at home and overseas, Vette became increasingly persuaded that TE901's crew were "innocent victims, trapped by a unique web of circumstances." He appreciated that "the public had been soaked for a long time in the belief that Jim Collins had carried his crew 'yelling and kicking' down through cloud until they smashed into the mountain"—and hence entertained no illusions of what might happen to him personally in promoting a hypothesis antithetical to the orthodoxy of the Chippindale Report. Vette would be going up against "old friends and colleagues of many years' standing, as well as the Chief Inspector, the Director General of Civil Aviation, and probably the airline's management and board of directors, to say nothing of the political considerations of disagreeing with the ultimate Head of Civil Aviation and chief shareholder of Air New Zealand—the Prime Minister." On the other hand, he was convinced that if a pilot of Jim Collins's caliber could be ambushed on the last leg of a trip to Antarctica, he would be grossly remiss not to make every effort to find out what *had* destroyed the outstanding captain's aircraft and all inside it.

Captain Vette pressed ahead with his research.

When the government's internal accident report was released, its practical effect was to absolve everyone at the airline of any responsibility for TE901's horrific crash in Antarctica. Subsequent media coverage sensationalized the chief inspector's findings with headlines such as "CRASH REPORT POINTS TO ERROR BY DC-10 CAPTAIN," "FLIGHT THOUSANDS OF FEET TOO LOW," "CREW UNCERTAIN AS PLANE NEARED SLOPE." By such means, Chippindale's message was spread far and wide in easy-to-grasp language: the gruesome disaster was the pilots' fault.

Once the Royal Commission hearings got underway in July 1980, Justice Mahon did not consider the Chippindale Report as potentially posing serious problems for him because he anticipated having little difficulty endorsing its author's findings. At least that is what he tactfully would recall in *Verdict on Erebus,* his best-selling popular account of the Royal Commission's investigation. Still, two things bothered him about the Chippindale Report. The first was that its author believed the undisclosed, last-minute alteration in the destination coordinates of TE901's nav track "had not misled the aircrew." Mahon was aware of newspaper reports stating that some Air New Zealand line pilots were attributing the crash precisely to the undisclosed coordinate change. The second issue was the obvious one of how a supremely experienced aircrew could have deliberately flown into or toward cloud, which Chief Inspector Chippindale defined as into or toward "an area of poor surface and horizon definition."

When his daughter was young, Mahon had once driven her past the High Court building while simultaneously reciting what he considered the "finest passage" of *The Divine Comedy:* the inscription Dante reads on the Inferno's gate. In his famous letter to Can Grande della Scala, Dante had declared some centuries earlier that his work's subject, if taken allegorically, is how man merits rewarding or punishing justice because of his freely made moral choices. For Mahon, the inscription's middle lines "refer to the concept that the entrance to Hell, and the inferno itself, was made by the Great One through his love and sense of justice to all mankind, which necessarily required that mortal sinners must forfeit their claim to paradise." Nowhere more tellingly than here, quoting Dante to his daughter in front of the High Court building, did the exceedingly guarded Justice Mahon make clear that the law was his temple.

Now, as the royal commissioner presided over extensive public hearings into the Erebus tragedy, he was becoming increasingly annoyed to hear certain repetitive mantras from the airline's unforthcoming executives day after day, month after month. Despite abundant testimony by line pilots to the contrary, nobody in senior management acknowledged knowing anything about a significant change in mid-1978 to the final stretch of the Antarctic charters' outbound route. From that date forward, the preprogrammed flight path had been switched from over an active volcano to over the flat sea ice of McMurdo Sound nearly thirty miles away—and the top brass were not aware of it over a year later? Equally odd, given the tremendous amount of publicity surrounding the popular practice of low flying, no airline executives appeared to be familiar with it. The protocol was perfectly safe because McMurdo Station's American air traffic control officers had agreed to monitor and guide the DC-10s to keep them clear of incoming military aircraft and local helicopter traffic in the area. The upshot was that while Justice Mahon intended his inquiry to be a neutral, fact-based search for the truth of what had sent a planeload of tourists on a deadly one-way trip to Antarctica, he could not extract any substantive information from the senior airline witnesses.

And "it was not merely the evidence given but the way it was given," Justice Mahon later said, which would eventually lead him to disbelieve what airline management had to say. By the very nature of their profession, judges are skilled at interpreting nonverbal cues to help them ascertain the credibility of those testifying before them. According to a professional with this type of specialized expertise, any number of these nonverbal signs are quite conspicuous in the televised segments of the Royal Commission's public proceedings.

In hindsight, it is apparent that much confusion and bitterness was generated by the government's decision to publish its own expert's analysis of the accident before the independent inquirer could even begin his own in-depth investigation. Prime Minister Muldoon and Air New Zealand CEO Davis had never wanted an independent inquiry at all but had lost that battle to Attorney General McLay. The best these forces could do was stall the start of the second inquiry long enough to get their own version of events widely disseminated first, do what they could to see that public opinion was shaped accordingly, and coach senior executives at the airline and its regulator how to stonewall when compelled to testify during the second inquiry.

As the Royal Commission hearings steadfastly proceeded, bad went to worse when captains John Wilson and Ross Johnson audaciously testified that, at their joint briefing session on November 9, they had made it crystal clear exactly where the computer track would take the five attendees. It was into Lewis Bay and across Ross Island, requiring the pilots to maintain a sixteen-thousand-foot minimum safety altitude until on Mount Erebus's far side. Being dead, two of those present at the briefing were unable to respond. Being alive and possessing some backbone, the other three attendees took issue with the briefing officers' claims when in the witness box.

One of that trio happened to be seated next to NZALPA's Captain Arthur Cooper during the presentation of Captain Ross Johnson, flight manager (line operations), who had run the simulator exercise at the joint briefing. As Johnson emphatically described to the court what he'd told and showed the five airmen concerning the route over Erebus, the erstwhile first officer Gabriel, now a captain himself, found himself repeatedly whispering to Cooper, "That never happened. That just didn't happen." When it was this pilot's turn to offer testimony under oath at the Royal Commission, he firmly denied ever being told the flight path would go over Erebus.

Captain Les Simpson, commander of the second Antarctic flight of 1979, was also one of the five attendees at the November

9 briefing. Taking exception to the "extreme fine detail [Captain Wilson] appears to have included for the [benefit of the] Commission," he found the embellished claims of both Wilson and Johnson so at variance with what he'd heard them say originally that he could scarcely believe the briefing they were now describing was the same one he had attended. By holding his ground calmly under increasingly hostile questioning, Captain Simpson's testimony cast doubt on the veracity of Wilson and Johnson's claims. In fact, all seven surviving pilots of the eight Antarctic charters flown in 1978–1979 would testify that the route on which they'd been briefed was not over Mount Erebus.

Why would Air New Zealand have created an Antarctic flight path that—whether for planning purposes or actual use—intentionally overflew an active volcano? And why would the carrier's regulator ever have approved such a route? The more the royal commissioner investigated the matter, the more apparent it became that "the most consummate confusion" had attended the planning of the airline's polar aerial tours. Partly this might have been owing to hastiness induced by competitive pressures emanating from Australia; Qantas was about to launch tourist trips of its own to the white continent. One sign of Air New Zealand's dysfunction at this early stage was its attempt to make the South Magnetic Pole the outbound destination for its first two charters. The airline had to be told that the South Magnetic Pole had wandered out to sea.

Curiously, it gradually emerged that Captain Gemmell had surreptitiously usurped the authority of his boss, Captain Keesing, who as director of flight operations was originally in charge of designing an optimal Antarctic flight path. Envisioning low flying in the McMurdo area, Captain Keesing had been working with Civil Aviation on a two-thousand-foot terrain clearance at the time Captain Gemmell quietly obtruded himself into the mix. The latter somehow managed to formalize with the airline's regulator both a track over Mount Erebus and the sixteen-thousand-foot minimum safe altitude regulation for pilots when overflying it.

Captain Gemmell's reason for cutting a deal so at odds with his direct superior's own negotiations-in-progress with Civil Aviation (and common sense) remains a mystery. Damning indictments of the potentially hazardous route would be made later by, among others, a four-time flight commentator on the aerial tours who, like Peter Mulgrew, possessed vast Antarctic experience. "I fail to understand why this route was chosen by Air New Zealand rather than the normal operational route," commentator Bob Thompson would testify, referring to the route up McMurdo Sound. He himself always advised the captain in command of each flight he was on "to follow the operational route from the northern coastline of Antarctica to McMurdo Sound rather than the route recommended by Air New Zealand." And they all did.

Any possibility of learning from Captain Gemmell himself the rationale for the Antarctic flights' parameters he had created died with him in 2012. Still, tantalizing clues do exist regarding his thinking as, behind the scenes, he finalized on paper Air New Zealand's flight path and minimum safe altitude for pilots headed to Antarctica. That he was allowed to become an Air New Zealand pilot without ever having served in the Royal New Zealand Air Force, which was the airline's normal recruiting pool, is suggestive. Perhaps this biographical detail helps to explain why Captain Gemmell never consulted RNZAF despite its decades of valuable flying experience in Antarctica.

Despite his family's extremely modest means, Captain Gemmell's single-minded drive from childhood to succeed in the aviation industry is also suggestive. It is difficult to imagine that a boy without resources who rose to become Air New Zealand's flight manager (technical) would not have felt perfectly capable of designing an exotic tourist excursion to a dangerous part of the world with minimal input from those foreigners controlling the skies in and around its outbound terminus. The American navy had readily approved Air New Zealand's proposed Antarctic flights in principle but had never requested any particulars. Captain Gemmell and any planning associates of his, for their part, never consulted the Americans for advice on the best approach to Hut Point Peninsula. They did not even apprise Mac Center that they had set an official route over Mount Erebus and a minimum safe altitude of sixteen thousand feet. Perhaps that was just as well because had they known the route and altitude parameters Captain Gemmell was putting in place, the Americans would most strenuously have objected to them. They did not allow even military aircraft to overfly Mount Erebus and would never have countenanced civilian aircraft doing so.

Later, Captain Gemmell created a secondary six-thousand-feet supplemental minimum safe altitude in a circumscribed sector near the tourist flights' outbound terminus. He did not inform American air traffic control about it either. The officer that served in 1979–1980 as McMurdo Station's chief traffic controller would later describe it in an affidavit for the Royal Commission as "absurd" given the limitations of local radar.

In 1977, when the airline's first six flights to Antarctica took place, they did carry a nav track that directed an aircraft straight at the mountain at the end of its outbound journey. However, *no sightseeing DC-10s ever overflew Mount Erebus.* Via their preflight briefing sessions, pilots commanding aerial tours in that first year were advised that they were free to disengage from the nav track once past the penultimate waypoint and fly visually up the sound. It was easy to do by simply deviating laterally from the route on New Zealand's books and swooping up the sound's flat sea ice parallel to the Victoria Land coast. From there the Americans' air traffic control officers could guide them on the final approach to the cluster of scenic attractions to be found on Cape Armitage, the tip of Ross Island's Hut Point Peninsula.

Among those who disengaged from the nav track at Cape Hallett was Captain Peter Grundy, a close associate of Captain Gemmell (and by 1978, his boss) who had evidently worked closely with him on establishing the parameters of the Antarctic flights. Captain Grundy was the pilot in command of the second aerial tour. Instead of entering Lewis Bay and passing over a volcano constantly emitting a monumental plume of steam and gas, he circumspectly shifted his route horizontally to fly along the Victoria Land coast before turning left to show his planeload of tourists the sights on Hut Point Peninsula's tip.

On his own tour in late 1977—it carried as an illustrious passenger the president of a division of McDonnell Douglas Corporation, the aircraft's manufacturer—Captain Vette likewise abandoned the nav track at Cape Hallett to follow the Victoria Land coast. To give his tourists a memorable view of Mount Erebus, which was visible for around 150 miles, he descended to 16,000 feet and flew up McMurdo Sound past, not over, the volcano. On approach to the aircraft's outbound terminus, local air traffic control cleared him to descend "to about 3,000 feet," then "circuit height of 1,500 feet." Captain Vette's distinguished guest, John C. Brizendine, was a former pilot himself. At the flight's conclusion, he congratulated Captain Vette and even wrote a highly complimentary article that later appeared in an aviation magazine and was distributed by Air New Zealand to every family in the country.

Of the six Antarctic flights of 1977, whose automatic navigation systems all held a route over Mount Erebus, even the behind-the-scenes designer of the Antarctic flights, Captain Gemmell, did not swear under oath at the Royal Commission hearings that he had taken his planeload of exuberant day-trippers over the mountain. What he reported was that after "deviating slightly whilst approaching Ross Island," he had conducted the last part of his outbound flight at the minimum safe altitude of sixteen thousand feet called for when overflying Mount Erebus. Did Captain Gemmell hope to insinuate that he *had* overflown the mountain when, in fact, his aircraft had entered McMurdo Sound just like all the subsequent flights (except that of Captain Collins)?

Challenged by Justice Mahon about the wisdom of designing, even if just on paper, a flight track taking tourists over an active volcano, Captain Gemmell serenely replied that it was perfectly safe so long as the mountain was not in eruption. If it were, he expressed confidence that someone would notify him beforehand. (British Airways Flight 9's near demise en route from Kuala Lumpur to Perth upon flying into a dense cloud of volcanic ash, about which no one had cautioned the crew, would not occur for another two years.) The fact that a thick wall of steam and gas continuously streamed some five thousand feet into the air from the volcano's crater did not seem to concern Captain Gemmell in the least. (It did concern Captain Vette, who "would not have overflown the plume of the volcano because of the discomfort through the air conditioning system and because of the possible damage to various types of filters.") Captain Gemmell also proved insensible to the danger of an aircraft being unable to communicate with the US Navy's radar and VHF radio installation, the Ice Tower, until safely on the far side of Mount Erebus. The temporary communications blackout at the vital endpoint of the outbound journey, which would last for forty miles, gave him not a whiff of unease.

It should have.

Whatever the rationale for originally establishing an Antarctic route that—on paper, anyway—involved crossing Ross Island and overflying Erebus, that route was superseded in mid-1978, when Air New Zealand acquired a new ground computer. It was to hold a digital flight plan for each of the airline's routes. The carrier's chief navigator, Brian Hewitt, personally input the one for Antarctica. In so doing, the destination waypoint's longitudinal coordinate was altered from 166 degrees east to 164 degrees east. The effect was to shift the aircraft's track at its southern terminus twenty-seven miles to the west of Mount Erebus. The computerized navigation track of future Air New Zealand Antarctic charters would steer their plane-loads of beguiled tourists up the perfectly flat terrain of McMurdo Sound to a new destination waypoint at the Dailey Islands.

The question as to whether the switch of the Antarctic flight path from over an active volcano to over flat sea ice was deliberate or accidental could not initially be established. Because he knew and had worked with him in the old days, Captain Arthur Cooper thought chief navigator Hewitt had honestly made a mistake—in Cooper's experience, Hewitt made mistakes. Before aircraft began flying computerized routes, they carried individual navigators, and Hewitt had been among their number. Whenever Hewitt was serving in that capacity on flights commanded by Cooper, the latter would recall, "I'd always keep an eye on him just because I didn't have that confidence in him." The man simply wasn't a natural navigator. He could "get from A to B," but it was a struggle.

After individual navigators were phased out, Brian Hewitt endeavored to make himself indispensable to Air New Zealand by becoming an expert on the computer systems that replaced them. He did not excel in his new specialty. Captain Cooper would say in an interview years later, "When he made the mistake from 166 to 164, I could quite believe it was Brian. And he'd set it up so that there was no cross-check of it." What Cooper meant was that chief navigator Hewitt had not double-checked his figures for accuracy for a destination waypoint located in perilous proximity to mountains on a forbiddingly frozen continent over 2,500 miles away.

Evidence that Air New Zealand did not make available until *after the publication of the Royal Commission's report* would conclusively establish that chief navigator Hewitt had indeed unwittingly made an error—two, actually—when programming the Antarctic flight plan into the ground computer during the summer of 1978. In the absence of that evidence, Justice Mahon tended to think the shift of the destination waypoint westward was probably deliberate. There were obvious safety-related advantages: the McMurdo Sound route ran over totally flat sea ice on roughly the same track used by all military aircraft servicing the American and New Zealand scientific installations, and it put Air New Zealand commercial aircraft at the critical end of their outbound journey in line-of-sight radar and VHF radio contact with the Ice Tower.

Although chief navigator Hewitt had made errors of which he himself was ignorant, it was clear that airline *management* knew from the outset about the changed destination waypoint and approved it in practice (even if no request was made to Civil Aviation officially to put the new route on the books). The proof was that suddenly Air New Zealand had begun churning out glossy new promotional literature, which included expensive passenger brochures. They contained a map showing a route up the middle of McMurdo Sound, and that map had been created with reference to a new track and distance chart that had sprung into existence contemporaneously with the revised flight path. The upshot was that senior airline personnel, if not chief navigator Hewitt, clearly knew in the summer of 1978 that the Antarctic flight path was no longer over Erebus (even on paper).

Pilots who flew the 1978 season's Antarctic aerial tours brought the new chart, which became known as Exhibit 164, to the Royal Commission's attention, explaining that they received it as part of their briefing materials. The airline itself, which had not disclosed the existence of this track and distance chart, was caught badly off guard. Exhibit 164 demonstrated that the revised flight plan ran up McMurdo Sound and did not overfly Ross Island. It made a mockery of Chief Inspector Chippindale's contention that the airline's Antarctic flight path had *always* gone over Mount Erebus.

According to Chief Inspector Chippindale, he had acquired his knowledge of the flight path from his closest advisor, Captain Gemmell, the most conflicted individual in the entire Erebus affair. He had asked what documents the accident crew would have been shown at their November 9 briefing and what else they would have acquired at Flight Dispatch in their Antarctic envelope before boarding the accident aircraft. Captain Gemmell had immediately produced a copy of a track and distance diagram from *late 1977* that indicated a route over Mount Erebus. Chief Inspector Chippindale claimed that Captain Gemmell had recovered this document at the crash site.

Considering what became labeled as Annex J to be a legitimate document the pilots had been shown at their briefing session and carried with them on the fatal flight served as a splendid pretext for the chief inspector to berate Captain Collins for imperiling his aircraft and endangering his passengers by approaching the mountain (he knew was there) at low altitude. Evidence presented at the Royal Commission established, however, that the pilots had *not* been shown Annex J at their briefing. It was Justice Mahon's conclusion on the balance of the probabilities that this document was *not* part of the 1978 or 1979 flight documents given to Antarctic crews.

And yet, there before Justice Mahon sat Air New Zealand's most prominent officials, all stonily insisting under oath that for a full fourteen months they had been wholly unaware of the Dailey Islands waypoint. Once it belatedly came to their attention in November 1979, of course, they'd quickly arranged to relocate that final waypoint back to a position directly behind Mount Erebus. They had thus "corrected" what the airline's CEO liked misleadingly to refer to as the "error."

Between management's professed lack of knowledge of the actual operating altitudes in McMurdo (the low flying) and its purported ignorance of the actual Antarctic route pilots were flying during the fourteen months leading up to the fatal flight (McMurdo Sound), considerable damage was being done to Air New Zealand's credibility and public image. Two concerned representatives of the New Zealand Air Line Pilots' Association (NZALPA), Arthur Cooper and Peter Rhodes, now sought to warn CEO Morrie Davis that Air New Zealand was being made a total laughingstock at the hearings. Although they managed to secure an interview with their boss, neither could say "a bloody thing" since he spent virtually the entire forty-minute meeting delivering a monologue on points he wanted *them* to appreciate.

———————

While Justice Mahon pressed ahead with the commission's hearings in the latter half of 1980, Captain Vette was making significant progress with his own research into the psychology of sight. He'd been focusing on two specialized fields of knowledge, one of which concerned attributes of a "mental set," the other of which involved the mechanics of "visual perception and deception." Roughly speaking, a mental set refers to the fact that, unlike a camera, human visual perception is a product of both eyes and brain and, as such, is influenced by what the viewer expects to see. Firmly instilled and repeatedly reinforced, such expectations constitute a mental set.

In the case of Captain Collins and his copilot, they had attended a preflight briefing, at which the route up McMurdo Sound had been both discussed and demonstrated—and the destination waypoint's specific coordinates shared. The ops flash line on the flight plan provided to them at Flight Dispatch prior to boarding was left blank, indicating that there had been no changes to that plan. The pilots carried with them to Antarctica a giant topographical map on which Collins, just the evening before, had plotted the aircraft's route over the sound's flat sea ice. And as they flew along on their demonstrably unerring nav track, they were visually confirming their position via specific terrestrial features that they expected to see.

Significantly, the pilots and Mulgrew had no difficulty identifying their arrival at McMurdo Sound's entrance from the coastline's contours and low line of black rock on each side of it. By an unlucky coincidence, Vette discovered, "the area they flew into [Lewis Bay] bore an uncanny resemblance to the area they believed they were approaching [McMurdo Sound]," given the weather conditions at the time. In short, the crew's mindset was that that they'd be flying up the sound, and nothing occurred until about a minute before the flight's grisly end to shake their belief that's where they were. In a manner of speaking, they'd been programmed to die.

Despite the crew's misapprehension about their location—which in fact was not McMurdo Sound but Lewis Bay—the DC-10 would not have collided with Mount Erebus had it not flown into a distinctive form of whiteout, about which neither Air New Zealand nor Civil Aviation considered it important to brief pilots operating in Antarctica. Vette's investigations into polar visual deceptions, in particular the insidious type termed sector whiteout, demonstrated how fallacious it was for the chief inspector to regard as negligent pilots untrained in sector whiteout who, once in it, failed to detect that they *were* in it.

According to Vette, this potentially deadly optical phenomenon occurs when "only a narrow sector of the total horizon is in whiteout." An aircrew can be carefully monitoring "atmospheric transparency" in all compass directions by reference to contrasting terrain, or texture, and nonetheless mistakenly conclude that they have a totally clear horizon rather than a partially compromised one. The danger arises when the sector in *unrecognized* whiteout lies directly in front of the aircraft, for this impairs visibility in the only direction that really matters.

Since passenger photos had demonstrated that visibility was virtually unlimited to the east and west of the aircraft, Captain Vette concentrated his efforts on ascertaining what combination of weather conditions would produce an extreme optical illusion in which an entire mountain lying directly in its path would be undetectable by the pilots, flight engineers, and even an Antarctic expert like commentator Mulgrew. The recipe proved to be this: a uniformly white terrain (no visual texture or contrast whatsoever), a solid cloud base above the aircraft with a pale underside (flying under an overcast), and sun shining from behind the aircraft at a thirty-eight degree or less angle from the horizon (which eliminates shadows). Under these circumstances, as Justice Mahon summarized Captain Vette's findings, "the fact that the forward vista of snow-covered terrain was rising to meet the cloud would not be observable and the intersection of the cloud base with the snow-covered terrain would appear to be a horizon located in the far distance." In other words, those on TE901's flight deck midday on November 28 were about to become victims of the "flat light illusion," by means of which they would mistake a snow *slope* for a flat snow *plain*.

In testimony that greatly irritated Air New Zealand's management since it had never briefed pilots flying the Antarctic route on the perils of this type of whiteout, NZALPA accident investigator Rhodes stated that his work in and around the crash scene had given him a sobering appreciation of the dangers of flying over snow under an overcast—that is, below cloud. "The matt surface of the snow gives *no depth perception*," he explained at the Royal Commission hearings, "even in conditions of fifty miles' visibility, and causes the wall of snow ahead to appear as a flat plateau with a distant horizon." Because of their "mental set" conditioning, TE901's crew members were anticipating a dash up McMurdo Sound over precisely the sort of flat plateau with virtually unlimited visibility described by Rhodes.

During his eight days in the witness box, Chief Inspector Chippindale was asked to explain why Air New Zealand's failure to brief TE901's pilots on whiteout shouldn't be considered a contributing cause of the tragedy. Even if one accepted his definition of probable cause—by which he curiously meant the accident's *immediate* cause—that did not preclude there being one or more contributing causes that preceded the immediate cause. In the present instance, the airliner had approached Ross Island from the north when, according to the Chippindale Report, surface visibility was good, surface definition and horizon definition were poor, and the mountaintops were cloud covered. Chief Inspector Chippindale had himself identified whiteout's hazardous hallmarks when flying: a loss of horizon, distance, and height perception, along with disorientation and a loss of references. As if by way of illustration, he had even described in his report the experience of a helicopter crew that landed on Ross Island's northern shore about fifteen minutes after TE901's crash. Because of their familiarity with the area, the helicopter's occupants knew for a certainty that Mount Erebus lay to the south even though they could not actually see any rising terrain in that direction.

Like all previous Air New Zealand pilots flying the Antarctic route, Captain Collins and first officer Cassin had not been briefed on the insidious ways in which polar whiteout swiftly degrades human visual perception, making it highly unreliable—particularly at low altitudes. The company's pilots were all trained in temperate climate flying and had no experience with, much less expertise in, white surface flying. Chief Inspector Chippindale did concede during questioning that over Ross Island the day of the accident was weather with a high potential for whiteout. Since the CVR made no mention of Mount Erebus, it seemed possible the omission might indicate that the aircrew had failed to perceive the rising terrain in their path—just like the helicopter crew cited by Chief Inspector Chippindale himself. The chief inspector was having none of it. Despite his theoretical knowledge concerning whiteout, he seemed not to grasp (or perhaps he merely *pretended* not to grasp) the ultimate implication of Captain Collins's decision to fly visually: he would be relying on direct information obtained through his *eyes.*

The problem with relying on information obtained exclusively through human eyes is that what is seen is not invariably what is there. An ordinary example would be a straight branch that only appears bent when extended under water. Encountering a dangerous polar optical deception for the first time, especially with no training in varieties of whiteout, a pilot would naturally believe himself to be perceiving all that was there to be detected by *human* vision—and he would be right. What Captain Collins and his copilot saw was a clear horizon. They had not been remiss in carefully monitoring their environment as they approached the end of their outbound journey: they *did* see all that could be seen by *human* eyes. Owing to sector whiteout on the north side of Mount Erebus on the afternoon of November 28, 1979, what the pilots saw was *not* what was right in front of them—a towering volcano.

Something new was perplexing Captain Vette as he continued his research into the tricks polar light can play. Despite his extensive experience in Antarctica, commentator Mulgrew had nonetheless committed a fatal mistake as the aircraft approached Lewis Bay. He had erroneously identified the capes to the left and right as being those at the entry to McMurdo Sound. How could that have happened?

Surprisingly, Vette discovered, it had proved easy to do. By a cruel trick of nature, the azimuth, or line, from McMurdo Sound to Cape Royds closely matched that from Lewis Bay to Cape Tennyson in both angle and distance. Additionally, the line from McMurdo Sound to Cape Bernacchi's cliffs matched that from Lewis Bay to Cape Bird's cliffs in angle, although the distance to Cape Bernacchi was three times longer than the distance to Cape Bird. By another cruel trick of nature, however, the latter's cliffs were only one-third as high as the former's. As a result, Gordon Vette concluded "they would subtend exactly the same retinal angle." The conclusion to be drawn from these findings was inescapable—they would have confirmed for Captain Collins and first officer Cassin that the aircraft was flying into the middle of McMurdo Sound.

Captain Ginsberg, an international authority on illusory visual phenomena, reviewed all of Captain Vette's data bearing on what the pilots and commentator Mulgrew, an Antarctic expert, were likely to have seen at the conclusion of the aircraft's second orbit. Summarizing his conclusions, Justice Mahon stated that "the two thin strips of dark rock to the left and right of the approach to Lewis Bay would coincide, in [their] opinion, with the entrance to McMurdo Sound." What's more, "if the Captain's Nav Track confirmed the pilot's belief that he was in the center of McMurdo Sound, then the totality of the illusion would be complete."

There had been, in fact, several conceivable opportunities for someone as familiar with Antarctica as commentator Mulgrew to realize that TE901 was not where it should have been. The first occurred as the aircraft passed over a speck of snow-covered rock outcrop thirteen miles north of Ross Island's Cape Bird. Beaufort

Island is seven square miles in size, with a high point of only 2,530 feet. Since there is no mention of this petite landmark on the aircraft's CVR, it appears that commentator Mulgrew did not see it.

Why not? When Beaufort Island hove into view of the photo-taking passengers, he himself is believed to have been making his way forward from the passenger cabin to the cockpit to resume his narration over the public address system. Upon arriving on the flight deck, he missed a second opportunity to see it because the aircraft was just then banking left at twenty-five degrees to complete the second of its two orbits in place. The tilt would have been enough to prevent him, once ensconced in his seat behind Captain Collins, from seeing Beaufort Island. It has been calculated that had commentator Mulgrew arrived on the flight deck a mere two minutes earlier, he would have spotted Beaufort Island down below to the right of the DC-10 and instantly realized something was terribly amiss. Beaufort Island was supposed to be on the left.

A final chance for Mulgrew to become aware of the aircraft's perilous location arose a moment later as the flight approached Cape Bird itself. However, by taking Cape Tennyson to be Cape Bird, he confirmed to the crew that the flight was exactly where it was supposed to be. As for Captain Collins, he found nothing about the forward view to be "off" until fifty-five seconds before impact. The DC-10 by then had reached a stage in its flight where, instead of an endless, flat white plain ahead, the pilots should have been able to see the scattered buildings of the Americans' McMurdo Station and the New Zealanders' Scott Base, as well as Mount Discovery beyond. In addition, although the aircraft was now, as Captain Collins thought, in the immediate vicinity of the radar pickup point, first officer Cassin could still not reach the Ice Tower on VHF radio. Although the aircraft's distance-to-run-indicator (HSI) showed TE901 now only twenty-six miles north of the final waypoint, Captain Collins judged it was time to fly away.

Chief Inspector Chippindale's analysis of the probable cause of what today still ranks as the worst aviation disaster in the Southern Hemisphere was repeatedly challenged at the Royal Commission hearings. Chippindale had attributed it to Captain Collins's decision, after descending below what the airline claimed was an inviolable minimum safe altitude of sixteen thousand feet, "to continue the flight at low level towards an area of poor surface and horizon definition." But wasn't it true that to *decide* to continue the flight toward an area with bad conditions, the captain would first need to see bad conditions up ahead? The chief inspector's academic grasp of whiteout by no means prevented him, in his report, from reprimanding Captain Collins for not actually seeing the optical illusion the whiteout over Ross Island had created.

The fact is that once Captain Collins perceived questionable conditions, he did not continue the flight at low level, as Chief Inspector Chippindale alleged. On the contrary, he did the precise opposite: he immediately decided to climb the DC-10 out of the situation. Chippindale chose to stand by his own interpretation of the doomed aircraft's critical last moments of flight. He appeared not to grasp that even if TE901 was, in fact, approaching an area of poor surface and horizon definition, Captain Collins needed to *perceive* it as such for his causation theory to hold. How else would he acquire the knowledge, through his eyes, that evasive action was urgently called for?

Chief Inspector Chippindale, unfortunately, was not the only professional involved in the Erebus case who refused to accept that sector whiteout creates visual deceptions which, by their very nature, those enveloped in them are incapable of detecting. Another was Captain Kippenberger, director of the Ministry of Transport's Civil Aviation Division, Air New Zealand's regulatory body. Even some members of the upper ranks of the judiciary professed themselves totally mystified. One of them was Sir Justice Owen

Woodhouse, president of the Court of Appeal, who would later have a very powerful, if malignant, role to play in the Erebus story.

As for Captain Collins himself, it was outside inputs, not perception of the rising terrain (since he could not see that), which compelled him to fly away. Justice Mahon would later sum up the familiar phenomenon of whiteout in the McMurdo area this way: "You have an ordinary overcast, and if you fly down McMurdo Sound at low altitude, it won't matter. If you fly into Lewis Bay at low altitude, then you are in mortal danger."

When Justice Mahon later retraced Captain Collins's exact flight track in an RNZAF C-130 Hercules, he could finally appreciate how the fatal optical deception had occurred. On his own approach to the crash site, he saw above it and to the left a rock outcrop. On the day of the disaster, though, the low cloud base had hidden it from view. Mahon also noticed a three-hundred-foot-tall ice cliff at the edge of Ross Island with two shallow strips of black rock running along its base. On the day of the disaster, however, the ice cliff was covered in fog, making the strips of rock at sea level indistinguishable from strips of plain old seawater. With no contrasts in terrain to enable depth perception, the crew would have believed they were low flying over a perfectly flat landscape, just as expected.

Mahon was led to a sorrowful reflection on the "inequalities of chance" after the wing commander of the Hercules in which he was riding—on the same path and at the same altitude and speed as the accident aircraft—rolled to the left out of the way of the rising terrain just twenty-five seconds before it would have collided with the mountain's lower slopes.

Suppose that there had been no discussion between Collins and Cassin about which way to turn.
Suppose that Collins, having decided to fly away,
had merely applied his power and climb settings,
pulled out the Heading Select knob, and turned it
left. His big aircraft, extremely maneuverable and
highly powered, would have rolled into a left turn
about 2,000 yards before the ice cliff and climbed
safely away to the east and then to the north, just
as we had done.

Captain Collins didn't do that because normal inflight procedures called for him to talk over the proposed change of plan
with first officer Cassin. Their conversation was brief—fifteen seconds—but the delay proved fatal.

When the ground proximity warning indicator came on six
seconds before impact, the crew probably believed it to be a false
alarm. As far as they knew, they were flying at low altitude over the
flat ice of McMurdo Sound. Nonetheless, they responded to the
"emergency" immediately by spooling up their three engines to 94
percent of maximum power and raising the nose up from the DC-
10's normal 5 degrees to 10.9 degrees, which enabled the airliner
to be climbing at ten feet per second when the 250-ton machine
hit the mountain and instantly disintegrated under an impact force
Vette cites as "200 times greater than that of gravity." The absurd
view of Captain Kippenberger, head of the airline's regulatory body,
that both Captain Collins and first officer Cassin must have been
simultaneously struck with a strange physical or mental malady
on the approach to Mount Erebus is belied by the consummate
professionalism the crew exhibited in their split-second response
to the ground proximity warning indicator.

APPROACH TO MCMURDO SOUND

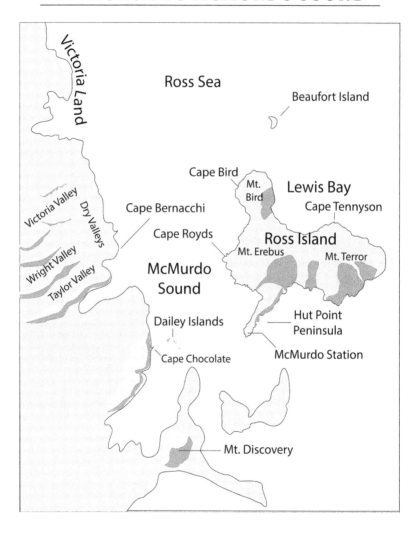

THE FOUR DESTINATION WAYPOINTS

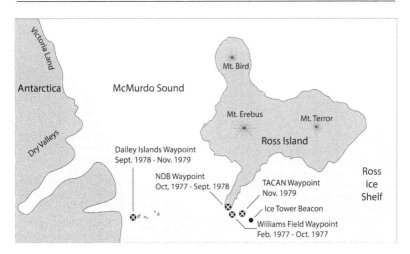

(1) **Williams Field** (166 degrees, 48 minutes east)
Over the ice runway associated with McMurdo Station, February 1977 (two flights)
> If used, it would direct an aircraft straight at Mt. Erebus.

(2) **The NDB** (166 degrees, 41 minutes east)
Over a beacon close to Williams Field, October-November 1977 (four flights)
> If used, it too would put an aircraft on a beeline to Mt. Erebus.

(3) **The Dailey Islands** (164 degrees, 48 minutes east)
Over volcanic islands in pancake-flat McMurdo Sound, 1978-1979 (seven flights)
> If used, it would direct an aircraft safely up the sound well to the west of Mt. Erebus.

(4) **The TACAN** (166 degrees, 58 minutes east)
Over a beacon close to Williams Field and the NDB, 28 November 1979 (accident flight)
> If used, it would direct an aircraft straight at Mt. Erebus.

10

THE HOVERING FATES

TWO WEEKS BEFORE CAPTAIN JIM COLLINS WOULD pilot the fourth Antarctic tour of November 1979, Captain Les Simpson commanded the second. As it happened, both airmen had attended the same briefing by Captains John Wilson and Ross Johnson in preparation for their respective trips, and both knew from that briefing that their route would lead up McMurdo Sound to the Dailey Islands. On his own flight south, Captain Simpson had disengaged from the navigation track at Cape Hallett and proceeded visually along the coast of Victoria Land on a heading leading to the Dry Valleys. Then, with American air traffic control's consent, he visually flew across McMurdo Sound toward the premier sightseeing area. Given clearance to descend to two thousand feet while overflying it, the captain and his copilot were now able to give passengers splendid opportunities to photograph at close range the sights in and around the tip of Hut Point Peninsula.

The maneuver of crossing McMurdo Sound from right to left involved taking the DC-10 from west of the navigation track across that track and flying eastward from there toward the scenic area at the flight's outbound terminus. In executing it, Captain Simpson was surprised to discover that the distance from the Dailey Islands waypoint to the TACAN, a beacon near McMurdo Station, was about twenty-seven miles rather than the ten or so he had guesstimated from the November 9 briefing. It wasn't a problem per se, but since destination waypoints were typically much closer to a beacon than twenty-seven miles, Captain Simpson thought future pilots of Antarctic aerial tours ought to be notified ahead of their flights of this wide "cross-track" difference. Accordingly, upon his return to Auckland, he phoned Captain Ross Johnson, flight manager (line operations), with this suggestion. Quickly, all hell broke loose.

After the crash of TE901 on the flanks of Mount Erebus, Captain Johnson initially claimed that "Captain Simpson rang me and said that the McMurdo position was in error and should be 166 degrees and 58 minutes," which was the TACAN's longitudinal coordinate. Later, at the Royal Commission hearings, he revised his recollection, now maintaining merely that Captain Simpson had opined that the destination waypoint would be more appropriately located at the TACAN than twenty-seven miles away from it. Captain Simpson strongly and consistently disputed this assertion, at one point testifying with convincing sincerity, "I did not report this matter to Captain Johnson as an error in position as I had no reason to believe the McMurdo position on the flight plan was other than a logical place to terminate the southern point of the flight plan track." Apparently, Captain Johnson had totally misconstrued the import of Captain Simpson's remarks. The misunderstanding would have fatal consequences.

Acting on what he took to be the point of Captain Simpson's phone call, Captain Johnson informed the Navigation Section's superintendent Lawton that there was an issue with the McMurdo position. It fell to him and chief navigator Hewitt to fix the presumed problem by relocating the destination waypoint to the TACAN. Incredibly, neither man then pulled up the airline's current Antarctic flight plan, stored in Air New Zealand's ground computer, to verify the destination waypoint's exact coordinates currently (before moving them). Questioned later during the Royal Commission hearings, Hewitt stated that he and his colleague had instead consulted earlier material on which the ground computer's Antarctic flight plan was based.

> Nicholson: The information which the captain of the aircraft, Captain Simpson, would have had, would have been the information on the flight plan which was issued?
>
> Hewitt: Yes.
>
> Nicholson: Would it have been possible to obtain a copy of that flight plan?
>
> Hewitt: It would have been possible, yes.
>
> Nicholson: Could you have got a printout of that flight plan on the computer terminal which was in your office?
>
> Hewitt: We could have done it if we'd gone about it a certain way.
>
> Nicholson: I suggest that it would have been appropriate and logical to have got a printout of the actual flight plan to compare with other material, would it not?

Hewitt: In hindsight, yes.

Nicholson: Can you suggest why the NV90 printout was obtained, rather than the flight plan printout?

Hewitt: No, we just went to the two sources of information on which the flight plan was based [the NV90 printout and Hewitt's own ALPHA worksheet].

Nicholson: *Rather than the actual flight plan itself?*

Hewitt: Yes.

Portentously, those two outdated sources of information concurred in giving the destination waypoint's longitudinal coordinate as 166 degrees, 48 minutes east, which was behind Mount Erebus. (It had become *164* degrees, 48 minutes east, out in McMurdo Sound, the moment Hewitt entered the flight plan into the new ground computer fully seven Antarctic charters prior to the one Captain Collins commanded.) When, therefore, Lawton and Hewitt compared what they erroneously took to be the current destination waypoint's longitudinal coordinate with the TACAN's—166 degrees, 58 minutes east—they discerned a discrepancy of only ten minutes of longitude, or just over two miles. So, Hewitt made the change.

In altering the destination waypoint's coordinates, chief navigator Hewitt would always insist the change was inconsequential, just a smidge over two miles from one location directly behind Mount Erebus to another nearby. Did the man have amnesia? The Antarctic route's destination waypoint for an extended period now had been a spot far to the west of the volcano. Chief navigator Hewitt himself had created the new track when entering the Antarctic flight path into the airline's new ground computer in July or August of 1978, over a year earlier. As a result, the current terminus of outbound flights was not some two miles (ten minutes of one degree of longitude) away from the TACAN—it was a full twenty-seven miles (a little over two degrees of longitude) away from it! Shifting the destination waypoint back twenty-seven miles to the east at this late date would have the effect of sending an aircraft approximately along the airline's original Antarctic route, which, at its southern terminus, led into Ross Island's Lewis Bay on a direct line with Mount Erebus.

There can be no doubt that, as a famously meticulous pilot, Captain Collins would have immediately studied the implications of the revised nav track if made aware of it. However, he was not. Airline management justified this lapse by maintaining that there was no point bothering Captain Collins with a trivial change in distance—which was all they claimed they were making. This does not make sense, however, as no one knew better than Captain Ross Johnson, who taught the simulator portion of the airline's Antarctic briefings, that the route ran straight up pancake-flat McMurdo Sound to the Dailey Islands waypoint a nontrivial distance away from the volcano. In a memo to the Navigation Section and two versions of his postaccident statement ("J2") to an Air New Zealand internal inquiry committee, he confirmed that Captain Simpson *had* reported the nav track was nearly thirty miles west of the TACAN. In any case, regardless of the size of a change to a flight plan, aviation protocol the world over requires the pilots flying the affected aircraft be apprised of it.

To Captain Cooper, who knew from experience that chief navigator Brian Hewitt tended to err, it was entirely conceivable that he had indeed made yet another programming mistake. Watching the man squirm while testifying—his nervousness was palpable— Captain Cooper concluded that Hewitt realized "exactly where the blame lay and who was responsible." The programmer, in Captain Cooper's estimation, had committed *both* computer mistakes— from 166 to 164 and then, over a year later, from 164 back to 166. In a final lapse, chief navigator Hewitt had not notified Captain Collins, who was about to command a flight carrying a computerized navigation track aimed directly at Mount Erebus. You just did not change a flight plan waypoint without informing the affected crew; it was utterly inconceivable. *You just did not do it.*

Originally, the revised Antarctic nav track was expected to go live on the aerial tour preceding that of Captain Collins. However, being unable to get the ground computer to accept the change in the flight plan in time for Captain White's journey, Brian Hewitt instead instructed the top officer at Flight Dispatch, out at the airport, to handwrite in the changed destination waypoint's coordinates on the master list given to the pilots for insertion into the aircraft's computer. That did not happen. Captain White therefore set out with his planeload of excited day-trippers in ignorance of the Navigation Section's foiled attempt to change the final leg of the nav track on his watch. Had Flight Dispatch hand amended Captain White's list of latitudinal and longitudinal coordinates for insertion into the onboard computer, word would have gotten back to Captain Collins, whose own flight was scheduled for the next week. Once aware of the altered coordinates, he was precisely the sort of pilot who would have swiftly investigated what it entailed.

Despite Captain Collins not being informed of the change in his computerized flight path—either indirectly in the manner just described or directly by an ops flash notation on his own master list of coordinates for insertion into the onboard computer the morning of his flight—his aircraft might yet have been saved. Air

New Zealand had gotten in the habit of forwarding to McMurdo Station's air traffic control facilities the DC-10's flight path for each successive "ultimate day trip." The information was conveyed in the customary waypoint format of latitudinal and longitudinal co-ordinates. Had the airline followed its own established protocol on the morning of November 28, 1979, the Americans would have detected instantly the changed track for the aircraft's approach to their air space later in the day and objected to it on grounds of safety. In the process, Captain Collins would have been alerted to the fact that he was flying on a nav track drastically different from the one he had been briefed on over two weeks earlier.

Curiously, though, on this occasion what Air New Zealand forwarded to Mac Center from Auckland was a series of latitudinal and longitudinal coordinates for every waypoint *except* the last one. To indicate the destination waypoint's location, an airline functionary needed to write 166 degrees 58 minutes east into a worksheet and add a symbol that ensured the changed coordinates would transfer over to the amended flight plan that would be typed into the ground computer and then, in abbreviated form, radioed to McMurdo the morning of the flight. However, the operator did not enter this symbol in the correct column of the worksheet. The result was that the flight plan radioed to McMurdo showed no destination waypoint coordinates—just the word "McMurdo."

Was this yet another improbable computer glitch or a deliberate attempt to mislead the Americans? Reportedly, the McMurdo air traffic control personnel had plotted the coordinates of the way-points of the initial 1979 flight and expected no changes for that year's series of tours. After the crash, the chief traffic controller and supervisor of Mac Center would tell Justice Mahon that the US Navy had no idea the flight track was *ever* anywhere other than McMurdo Sound. Had he known a civilian aircraft planned to overfly Mount Erebus, he would have forbidden it.

Whether by accident or design—Justice Mahon believed it was the latter—writing "McMurdo" instead of inserting the relevant coordinates ensured that local air traffic controllers would have no inkling that the airline intended for Captain Collins's aircraft suddenly to appear in their air space after popping over the top of a nearby active volcano. Mac Center and the Ice Tower fully expected Captain Collins's flight, like Captain White's of the previous week and Captain Simpson's of the week before that, to proceed up the customary track in the middle of the sound.

Even in ignorance of the altered track, Mac Center might still have been able to save TE901. The opportunity existed just prior to the crash when Captain Collins was executing the two leisurely figure-eight descents that would bring him down to two thousand feet for the dash up McMurdo Sound. During those orbits, the DC-10 briefly emerged from behind the mountain in a direct line of sight with McMurdo's radar facility, known as the Ice Tower. Stationed there were both a radar operator and a radio operator, the latter capable of communicating at line-of-sight frequencies (VHF and UHF). Nearby was Mac Center's massive communications complex, at which worked a radio operator with an HF transmitter, which is not dependent on line of sight. We know the aircraft's transponder received a radar signal from the Ice Tower. On the tower's radar screen, the DC-10's track would have appeared as a row of blips. If, seeing the blips, the radar operator had alerted the radio operator and the latter had managed to get a warning through to TE901, Captain Collins could have applied "fire wall" (maximum) power, turned, and flown safely away. However, Captain Collins received no such warning.

Whether anyone at McMurdo tried to send him one is, like much else involving the Erebus disaster, shrouded in mystery. On the one hand, the official who ran Mac Center believed the DC-10 had not been identified on the radar screen. He noted that it would have been visible only fleetingly. Furthermore, there was no reason for the radar operator to be keeping a close eye on his screen since TE901 was flying visually.

True enough. The only complicating factor was that Mac Center's master tape recording of all communications to and from both the Ice Tower and Mac Center had nothing at all on it in the last four minutes and forty-two seconds of TE901's flight. Initially it was said this was simply because no transmissions were initiated during that critical interval. However, NZALPA's air accident investigator Rhodes had made inquiries of his own at McMurdo Station and Scott Base right after the accident and been told the tape's last portion was accidentally erased.

Accidentally—or deliberately? If the latter, it could only have been to conceal some conceivably negligent act on the part of US Navy personnel. Although not allowed to interview the three operators on duty at the Ice Tower and Mac Center the day of the crash, Justice Mahon had no difficulty imagining a scenario under which it might have been thought advisable to "lose" the last several minutes of recorded tape. Assuming the Ice Tower's radar operator had seen the DC-10 on his screen, he would have exhorted the radio operator to warn TE901 it was to the north of Erebus and headed straight for it. That transmission necessarily would have been on line-of-sight VHF. But because the aircraft had quickly disappeared again behind the mountain after Captain Collins's orbiting sequences, no warning given only on VHF could possibly have reached it. Mac Center's own radio operator, however, who was tasked with monitoring all the Ice Tower's transmissions, would have overheard the dire warning from his colleague stationed there. Perhaps it did not occur to him to repeat it on HF, which the pilots *could* have heard.

Further light would be shed on this mystery several years later, in 1987. That was when the families of some of TE901's casualties filed a civil suit against the US Navy on the grounds of negligence. Erebus authority Stuart Macfarlane visited the California offices of the plaintiffs' attorneys after the trial was over (the case had been thrown out). There they shared their conviction that Justice Mahon's suspicion of tampering with the tape was correct. However, "the U.S. military put up a protective shield around the witnesses," and the plaintiffs' lawyers could no more penetrate it than Justice Mahon could during his earlier inquiries into the matter. Did this stonewalling mean the US military *did* have something to hide? There was talk of drug usage at the base, which suggested to Macfarlane that an operator "may have been incapacitated." The plaintiffs' attorneys at the US trial also told Macfarlane that the radar operator had suffered a mental breakdown, which is ominously suggestive.

— ◆ —

It is difficult not to share Justice Mahon's doleful sentiments concerning the role chance plays when one reflects on the multiple ways in which, but *for* chance, Captain Collins and first officer Cassin could have become aware that their aircraft was not where they believed it to be. Commentator Mulgrew might have arrived on the flight deck early enough to spot Beaufort Island on the wrong side of the aircraft, or he might have correctly identified Cape Bird instead of confusing it with a close facsimile in Lewis Bay. Either discovery would have set off massive alarm bells. Ross Island might not have been experiencing whiteout that afternoon, in which case all five men on the flight deck would have seen Mount Erebus up ahead and Captain Collins could have taken evasive action. Flight Dispatch might have alerted Captain White of the aircraft's changed destination waypoint, not yet able to go live in the ground computer, or Captain Collins might have been alerted by an ops flash notation on the top of his printout of the flight track's coordinates that a revised route had just gone live. Air New Zealand might have sent Mac Center all the flight's waypoints that morning in the conventional format and had its dangerously revised flight track quickly exposed by US Navy air traffic controllers. The airliner might have been detected, if only fleetingly, on the Ice Tower's radar by an attentive operator and a radio warning successfully conveyed just in the nick of time.

Chance inequalities did not end here. The undisclosed rerouting of the flight path back to its original position behind Mount Erebus had been scheduled to take effect on Captain White's flight but got delayed until Captain Collins's the next week. The change would not, however, have killed everyone on the former's aerial adventure if it had gone live a week earlier than it did without notifying *that* flight's commander. Why? Because Captain White disengaged his nav track seventy miles out from Hut Point Peninsula. According to US Navy personnel and Captain Vette as well, Air New Zealand pilots flying the Antarctic route regularly unlocked from the automatic navigation system after reaching Cape Hallett, the penultimate outbound waypoint. They then descended visually to take advantage of the best sightseeing opportunities in the McMurdo area. However, as cautious a pilot as he was known to be, Captain Collins had prudently decided to use his nav track as a backup system to flying visually. He therefore did not unlock from it except to execute his two descending orbits. That nav track, reprogrammed only hours before his flight with no notification given to him, then unerringly directed Captain Collins's aircraft straight into Mount Erebus.

11

THE ROYAL
COMMISSIONER'S
CONCLUSIONS ON
CAUSATION

FOR DECADES, JUSTICE MAHON HAD BEEN PASSION-
ate about the law, "his one truth." It must have greatly disturbed
him that, as he came to appreciate, a determined effort was underway by the Muldoon administration, the government's office
charged with investigating air accidents, the management of Air
New Zealand as well as its regulator, and a troupe of lawyers to
manipulate the New Zealand legal system to achieve their own
nakedly partisan ends. Sam Mahon would later distinctly recall
his father's complete rejection of the evidence given at the Royal
Commission hearings by witnesses that he knew, through his own
investigatory efforts, were lying. "In my life as a kid, if I had ever

taken that stance, I would have been in deep s***," he observed. "My father drew very, very severe lines [between truth and its opposite]." In his official report, Justice Mahon did not mince words; his substantive findings painted a picture of a seriously dysfunctional airline, not of incompetent pilots.

In a video interview conducted after the Royal Commission of Inquiry's conclusion, Justice Mahon would credit Gordon Vette with having "put me and counsel to the Royal Commission onto the right path." Being on Vette's evidentiary path enabled the judge to produce, in the Mahon Report, an analysis of the Erebus accident remarkable for its pioneering application of systemic analysis to the issue of aviation safety. The duo's intellectual breakthrough was to grasp that latent failures—that is, preexisting dysfunctions—in an airline's management and administrative sectors can result in catastrophe as surely as mechanical failures or pilot errors can. Such systemic factors within an airline's organizational setup Vette defined as "small acts, errors, or omissions" that could possibly be innocuous individually but that, "when combined, form a trap that defeats the pilots and the aircraft's detection and defense mechanisms." The result inevitably would be a crash.

It had been the resourceful researcher Captain Vette who identified the sector whiteout optical illusion that prevented the five men on TE901's flight deck from seeing the volcano, thereby clarifying for Justice Mahon the circumstances surrounding the accident's proximate cause while he himself probed its dominant cause. Although unable to pry Air New Zealand CEO Morrie Davis loose from "the flying in cloud, pilot error theory," Vette was a crusader, working tirelessly with the royal commissioner as a special adviser to identify not only the whiteout issue on the day of the fatal flight but also those organizational airline deficiencies that were ultimately responsible for the deaths of 257 people. A conspicuous one, he found, was the company's system for transferring vital information on the safety-related topic of white surface flying. While plenty was known about its perils, *nothing* had been communicated to those who needed the information most: Air New Zealand pilots commanding an Antarctic aerial tour.

Although Captain Vette had obtained permission from Morrie Davis to testify at the Royal Commission hearings as an individual, the airline's CEO was "later very angry about that." In fact, Air New Zealand rejected all suggestions from Mahon and Vette that the Flight Operations area of the company exhibited weaknesses with respect to flight safety—weaknesses that posed enough danger and risk that the airline's protocols and practices required an overhaul. In keeping with its hostile stance toward those who dared question the integrity of Air New Zealand's operations, management began to pressure Captain Vette to leave the airline in which he had built a highly successful flying career over decades of distinguished service. He resigned in June 1982, noting that "there were a lot more casualties out of Erebus than those who died on the mountain."

Despite the compelling ocular evidence submitted to the Royal Commission by Captain Gordon Vette, Captain Kippenberger of the Ministry of Transport's Civil Aviation Division could not be persuaded to abandon his conviction that there had been no whiteout problem on the day of the disaster. (This, it will be recalled, had led him injudiciously to account for TE901's crash by speculating that the pilots must have suddenly and simultaneously been afflicted by a mysterious illness.) In its final submissions to the court, the head of the national airline's regulatory body thus continued denying that any unusual visual deceptions could have occurred and persisted in championing the chief inspector's view that pilot error was the sole cause of the disaster.

The pilots' "error" had been to descend below the "official" minimum safe altitude of 16,000 feet on approach to McMurdo. It did not matter that, at the end of his extended testimony—which was made ten weeks into the Royal Commission hearings—briefing officer Captain John Wilson had dropped a bombshell. Up until then, this key airline executive had stressed that the Antarctic flight path always lay over the mountain, with the minimum safe altitude until on its far side being an inviolable 16,000 feet. Suddenly, in closing his testimony, Wilson startlingly revealed that in 1978 he'd become aware, via individuals in Flight Operations, that aircraft were descending below 6,000 feet at the discretion of McMurdo's American air traffic controllers. Apparently, Air New Zealand had given crews the authority to do this. It was a stunning admission even if Captain Kippenberger, the regulator's top man, disavowed any knowledge of the newly disclosed company policy—or of the well-publicized practice of the airline's Antarctic tourist flights themselves, for over two years now, to operate at US Navy approved low altitudes at their outbound terminus.

Whether "cling[ing] like a leech" (to use Justice Mahon's simile) to this position would shield Air New Zealand's regulator from liability for damages in future civil litigation was unclear. Before the Royal Commission hearings, few had imagined that Civil Aviation might have some real liability for the disaster. But at the hearings, it had become apparent that the regulator had not energetically monitored the airline's Antarctic operations in accordance with its statutory duty to do so. Instead, it had been functioning essentially as a rubber-stamp agency with respect to Air New Zealand's requests involving its tourist flights over a remote and forbidding continent.

Where was the regulator's safety calculus? Initially, Civil Aviation had deferred to the airline's proposal for a flight path over an active volcano that the three military services flying in the region for years considered unacceptably dangerous. It never insisted that Air New Zealand apprise the US Navy, which controlled the air space in the McMurdo area, of its envisioned flight track—much less seek approval for it. The regulator did not require Antarctic crews to be briefed on the various ways in which our primarily visual species can be tricked by the polar regions' commonly occurring ocular illusions. It did not instruct the airline to give aircrews a training session on landmarks in the greater McMurdo area and to provide one or more topographical maps of it. It didn't proactively ensure that Antarctic-headed captains first took a familiarization flight to that vast and malevolently white wilderness, as required by law. Worse, when after the fatal flight Civil Aviation became aware of Air New Zealand's long-standing violation of the requirement, it moved swiftly to grant the airline a dispensation from it.

The highly trained commanders of military aircraft operating in Antarctica knew the full panoply of challenges in flying over endless miles of hostile, frozen terrain. After TE901's crash, they expressed incredulity upon learning that New Zealand pilots in command of tourist flights to the white continent were not required, for safety's sake, to take even a single orientation trip there and back under supervision. The military personnel at McMurdo believed if Captain Collins had done so, he himself would have recognized Beaufort Island—on the wrong side of the DC-10—and realized the nav track must have been reprogrammed without his knowledge. There would have been ample time to take evasive action, thereby saving both the aircraft and its otherwise doomed occupants.

Civil Aviation's dismissive attitude toward the importance of disseminating information about polar optical illusions and the cruciality of prior Antarctic experience for Air New Zealand pilots heading to the white continent is difficult to fathom. Because of it, those commanding tourist flights to this dangerous part of the world had the benefit of neither. They'd *all* been put at risk, Captain Vette realized with a start after gaining expertise in the mechanics of human perception and the hazards of white surface flying.

Not being an airline operator, the Ministry of Transport's Civil Aviation Division was not covered by the Warsaw Convention like

Air New Zealand itself was. If negligence could be established against any of the division's officials, "in the sense of any culpable act or omission" that caused or contributed to the disaster, the Ministry of Transport would be exposed to claims for millions of dollars' worth of damages. The government's in-house accident investigator had providentially asserted that pilot error was the sole cause of the Erebus disaster. If that finding were to stand up in court proceedings, claims against the airline would be capped at a mere forty-two thousand dollars per claimant in accordance with the Warsaw Convention. At the same time, such a finding would render the Ministry of Transport impregnable from a claims standpoint. It is no wonder that Civil Aviation was at pains to hold, with the chief inspector, that pilot error had been the one and only factor in the crash of TE901.

Although the government's case for pilot error had steadily lost credibility as the Royal Commission hearings relentlessly proceeded, it remained for Air New Zealand, as for Civil Aviation, the ideal pretext under which to avoid any responsibility for the catastrophe. The airline thus vigorously continued to promote this attribution of blame in its closing submissions. Experienced as he was in the entire spectrum of cover-up tactics used in disaster cases, Justice Mahon was naturally familiar with the tendency of companies facing potential legal liability to produce corporate witnesses conspiring to tell a duplicitous tale. It did happen, and it was happening here. Still, the royal commissioner was stunned by how "unintelligent and obtuse" these particular Air New Zealand witnesses were; it was as if they'd not even tried to devise a compelling storyline. He thought management must have been prepared to "brazen the whole thing out to the end" since CEO Davis had let it be known that the government stood with the national airline.

In a provocative mental exercise, Mahon asked himself what would have happened if the undisclosed altered flight plan coordinates had produced no disaster—if instead the pilots had been able to figure out, in time, that something was wrong and climb safely away. "In due course there would have been instituted in New Zealand a public inquiry into the incident," Mahon thought. At *that* inquiry "the persons placed on the defensive from the outset would have been the relevant personnel of the Flight Operations Division of the airline."

Mahon could envision how such a proceeding would have played out.

> Captain Collins would have produced the whole of the contents of his flight bag, and they would have included his maps, his atlas, all his flight documents, and his black ring-binder notebook with all its pages intact. The crew would have testified as to the pre-descent briefing, and the pilots would have been able to say exactly what they saw on the approach to Ross Island. I doubt very much if there would have been too much heard at such an inquiry, with Captain Collins, First Officer Cassin, the two flight engineers and Peter Mulgrew present and listening, about wrongful reliance on the inertial navigation system, unlawful descent below minimum safe altitude, flying towards an area of deteriorating visibility, and the like.

> On the vital question of visibility there would have been, I need hardly say, the evidence not only of the flight crew but also of large numbers of passengers who must have looked at Ross Island in the course of the orbiting turns which the aircraft made.

Justice Mahon was well known for his ability to "cut through argument and humbug" to expose key issues in a case. Additionally, he
possessed the enviable skill of analyzing a great number of complex
issues in support of his conclusions regarding them. In keeping
with this style, the royal commissioner approached the determination of cause and culpability in the Erebus disaster by zeroing in
on the altitude evidence and the navigation evidence.

Regarding the altitude evidence, he was convinced that airline
briefing officers *did* authorize aircraft routinely to descend below
the "official" sixteen-thousand-foot minimum safe altitude when
approaching McMurdo. There had been the totally unexpected
disclosure of the practice by a senior airline executive, Route
Clearance officer John Wilson, late in his testimony. There was the
confirmatory courtroom testimony of captain after captain in command of an Antarctic charter in 1978–1979. Starting even earlier,
in 1977, there was already a steady stream of newspaper articles
favorably describing the popular practice of low flying at the end of
a charter's outbound journey. And, finally, there was the testimony
(or non-testimony) at the hearings of Air New Zealand's CEO
Davis himself. Justice Mahon had asked him how the airline—
without his personal knowledge—could have reproduced and mailed
out *one million* copies of an article in *Traveling Times* describing
the low flying. "He simply turned towards me and spread his arms
outwards in a despairing gesture. He was indicating his total lack
of comprehension that such a thing could have happened."

Reflecting on the principal navigation issue—why the accident
aircraft's track led toward Mount Erebus—Justice Mahon reached
several straightforward conclusions. The first was that in 1978 and
1979 the Antarctic route lay, "and was known by those concerned in
the company to lie," not over a mountain but up McMurdo Sound.
The second was that all crews in those years were told at briefing that
their route was up McMurdo Sound. The judge's third conclusion

was that mere hours before the fatal flight, the airline's Navigation Section had altered the route to lead over Ross Island on a direct line with Mount Erebus. His final conclusion was that, by an egregious blunder, no one had informed the pilots of that change.

In all, Justice Mahon identified ten contributing causes to the disaster, in the absence of any one of which it would not have occurred. However, he selected as the "dominant cause" the one that "continued to operate from before the aircraft left New Zealand until the time when it struck the slopes of Mt. Erebus." That was "the act of the airline in changing the computer track of the aircraft without telling the aircrew." Given the mountainous terrain to the left and right of pancake-flat McMurdo Sound, a twenty-seven-mile alteration to TE901's flight track was obviously consequential, at least potentially.

There was a highly conspicuous drawback to the stubborn insistence of Air New Zealand's executive witnesses that they knew absolutely nothing about (a) any low flying by the airline's pilots in Antarctica and (b) the existence of the McMurdo Sound flight path. For those assertions to be true, the Air New Zealand witnesses would have had to commit an ungodly long sequence of errors over a protracted period. These errors have been succinctly described by the outstanding Erebus researcher Stuart Macfarlane. There were a staggering number of them, Macfarlane found: 177 altitude mistakes and 54 navigation mistakes. For the airline's case of pilot error to be credible, a person would need to accept that Air New Zealand's aviation experts had committed *all* of them.

It was too much for the royal commissioner to swallow: "I was quite unable to accept all of these mistakes could possibly have been made." It was logically conceivable, of course, but highly improbable. Besides, Occam's razor (the law of parsimony) holds that a simple explanation is preferable to a convoluted one. The simple explanation required Air New Zealand not to have made over two hundred mistakes, just one—not to inform the pilots it had changed their aircraft's computerized flight path. This grievous lapse was not, however, Mahon stressed, a stand-alone phenomenon. It was symptomatic of shamefully careless safety-related administrative practices and widespread communications problems within and between different sections of the airline's Flight Operations Division. Perhaps these were to be expected in a company whose CEO had proudly evolved a management system that depended on verbal communications, eschewing written instructions, written protocols, and written reports.

In his original remit, Justice Mahon had been given ten points of reference for his inquiry, the most important of which was to establish whether any "culpable act" had led to the disaster. In what became known as the Mahon Report, released in April 1981, he found that Air New Zealand was riddled with substandard administrative, management, and organizational procedures. These glaring deficiencies he deemed responsible for the unconscionable loss of 257 lives on a remote mountainside. "By a navigational error [the coordinate change] for which the aircrew was not responsible, and about which they were uninformed," Justice Mahon elegiacally wrote, "an aircraft had flown into Lewis Bay, and there the elements of nature had so combined, at a fatal coincidence of time and place, to translate an administrative blunder in Auckland into an awesome disaster in Antarctica."

12

THE ROYAL COMMISSIONER'S CREDIBILITY FINDING

GIVEN THE GRAVITY OF THE CASE HE WAS INVESTI-gating, Justice Mahon had carefully weighed whether to ignore ordinary judicial protocol, which frowned on any decisive finding of perjury, preferring a judge simply to indicate his acceptance or rejection of a particular line of evidence when the testimony was in conflict. The Erebus disaster was, however, no minor tragedy; fully 257 people had lost their lives in TE901's horrific crash into a faraway mountainside. In paragraph 377 of the Mahon Report, the royal commissioner thus found it appropriate to ignore judicial convention.

Had the judge been intent on surviving in his chosen profession, he could have tempered his remarks on the credibility of the executives testifying before him over many vexing months at the

Royal Commission hearings. A more circumspect investigator might, for example, have said that he did not believe witnesses X, Y, and Z and left it at that. It was obvious to commentators after the Mahon Report came out that the language in paragraph 377 had been deliberately selected for its power and what they termed "pungency."

It was a singular feature of Mahon's investigative methodology that he based everything on the facts as he uncovered them or as they were introduced into evidence. To these he then applied his penetrating insights into human nature and the human condition. In his role as royal commissioner probing the unthinkable, the result was an astringent assessment of the motivations of the parties providing testimony. The Chippindale Report, Justice Mahon had come to appreciate, was plainly *not* a disinterested government investigator's good-faith analysis of the causal factors involved in TE901's crash. It was, rather, an energetic exercise in subterfuge designed to obscure Air New Zealand and Civil Aviation's responsibility for that debacle by attributing it instead to the aircraft's dead pilots.

In the judge's damning view, reinforcing the correctness of Chippindale's "pilot error" mantra had been the dishonorable objective of all senior Air New Zealand and Civil Aviation management personnel testifying before him at the extended Royal Commission hearings. His sensibility and patience sorely tried by their deceitful conduct over many months, Mahon memorably underscored these men's utter lack of credibility with words that have echoed down the decades.

> No judicial officer ever wishes to be compelled to say that he has listened to evidence which is false. He always prefers to say, as I hope the hundreds of judgments which I have written will illustrate,

that he cannot accept the relevant explanation, or that he prefers a contrary version set out in the evidence.

But in this case, the palpably false sections of evidence which I heard could not have been the result of mistake, or faulty recollection. They originated, I am compelled to say, in a pre-determined plan of deception. They were very clearly part of an attempt to conceal a series of disastrous administrative blunders and so, in regard to the particular items of evidence to which I have referred, I am forced reluctantly to say that I had to listen to an orchestrated litany of lies.

The royal commissioner also ordered the airline to pay NZ$150,000 toward the cost of the inquiry. He was empowered to do so by the terms of The Commissions of Inquiry Act 1908, which allows an inquirer to impose a cost order on any witness who unjustifiably prolonged, obstructed, or added undue cost to an inquiry. Air New Zealand's unforthcoming witnesses, he believed, had done precisely that.

Years after the words "an orchestrated litany of lies" were broadcast around the world, Margarita Mahon would explain the genesis of the sentence in which they are embedded. Peter was a wordsmith from childhood, she would recall, when he had already started collecting boxes of words and phrases. In the Erebus matter, he had consulted multiple dictionaries to ensure that his official report's credibility finding was phrased just so.

In Margarita's telling, the judge's irritation stemmed not so much from the Air New Zealand executives themselves but from the lawyers representing them at the Royal Commission hearings.

They could have presented "more than they were doing" and it "annoyed [Peter] because they knew better." From the judge's perspective, no doubt the principal offender in the category of lawyers that "knew better" would have been his erstwhile good friend Lloyd Brown. Brown was, after all, the lead lawyer representing Air New Zealand. He was also the individual on whose personal recommendation Justice Mahon had wound up being appointed royal commissioner. "They were taking me for a fool, which I'm not," Margarita reported her husband as declaring.

———◆———

Despite knowing that the accomplished Captain Collins had not been made aware of the last-minute change to TE901's nav track, Prime Minister Muldoon had ordered no deep probe into the character of the individual being proposed to lead an independent inquiry into the cause or causes of the aircraft's collision with Mount Erebus. Since the government was admittedly under growing pressure to appoint a royal commissioner, it is possible Muldoon felt he had no time thoroughly to vet Mahon. He may have counted on Auckland attorney Brown to keep his associate in line. However it happened, Prime Minister Muldoon's failure to investigate the would-be investigator was proving to have been a fatal omission because Mahon's bedrock values were discernable in his legal conduct and decisions over many years.

The most character revealing of his cases, perhaps, involved the 1950s Parker–Hulme "murder of the century." The cold-blooded killing by two girls of the mother of one of them was notorious—so notorious that future *Lord of the Rings* producer Peter Jackson would one day turn the story into a psychological thriller called *Heavenly Creatures*. It appears that no one in Prime Minister Muldoon's circle was aware of this episode, much less Mahon's pivotal role in it. By early 1980, however, when the government needed to select a royal commissioner, these events would have been very old news indeed.

Acclaimed New Zealand author James McNeish was one individual who did review the Parker–Hulme case and other legal proceedings involving Peter Mahon. The picture he eventually formed was of "a man of integrity who was nobody's lackey and who could not under any circumstances be bought." Justice Mahon might have embarked on his official inquiry into TE901's crash with the comfortable expectation that he would be rubber-stamping the government investigator's own findings in the fullness of time. That did not happen, though, because the new royal commissioner was soon beset by problems.

One issue proved to be a dearth rather than a wealth of airline documentation on its operations, especially with respect to the Antarctic charters. Another challenge was a roomful of unforthcoming—if not outright hostile—airline witnesses that, for months on end, gave testimony so incredible it strained credulity past the breaking point. Justice Mahon's suspicions, first aroused when Air New Zealand declined to make any opening statement of its position, grew as the Royal Commission hearings progressed. If Erebus was to prove Justice Mahon's Waterloo, it was also to become his capstone. Come what may, this singular man with his nerves of steel would not be dissuaded from doing his duty by the 257 individuals whose lives abruptly ended when the DC-10 in which they were sightseeing collided with an Antarctic volcano at 1:50 p.m. one fine day in November 1979.

His father was an "uncompromising adversary," Sam Mahon once observed. "I never saw him back down from anything." Even in a situation as perilous for him personally as rendering his honest legal opinion regarding culpability in the Erebus accident, the judge was prepared to act as conscience dictated. Unlike much of the rest of the country, Justice Mahon somehow was not affected by Prime Minister Muldoon's overbearing manner, abusive outbursts, or the general atmosphere of fear that he engendered.

The sensational findings contained in the Mahon Report were released on April 27, 1981, while Justice Mahon was out of town. After submitting his official report, he had retreated to the South Island to go duck shooting with his son. There, the judge predicted a stormy reception of his analysis by his establishment peers. He was not wrong, for a huge political conflagration instantly ignited, the ferocity of which New Zealand had never before experienced. Overnight, the royal commissioner had single-handedly turned the tables on the national airline by absolving the pilots of any responsibility for the terrible disaster in Antarctica. The responsibility, he claimed, lay with Air New Zealand itself—specifically, with its dangerously dysfunctional internal organizational arrangements.

For the shocked public, which may have naively assumed that gross corporate malfeasance and cover-up conspiracies occurred only in other countries, Air New Zealand's CEO, Morrie Davis, would soon become a folk villain. Assuming the mantle of a folk hero, Justice Mahon would shortly start finding flowers on his home's front doorstep and strangers picking up his tab upon recognizing him in a restaurant. The public's adulation of Justice Mahon would continue indefinitely. Nonetheless, an extremely ugly political slugfest was about to start in the corridors of power.

Looking back decades later, his close colleague and great friend Ted Thomas would suggest that Mahon's supreme gifts as a writer may have tripped him up. "This might have been one of those occasions," he said, when the temptation to deploy a theatrical phrase "overcame prudence," a point made at the time of the Mahon Report's dissemination by Prime Minister Muldoon himself. Thomas was not satisfied that Mahon wasn't correct in his finding; he "just" regretted the forceful language his friend had used to express it. Had Mahon only conferred with others on the wording of his proposed credibility finding, "any lawyer" could have told him that what he wanted to say to the media "could have been said in a much, much better way." Without that single literary flourish, in short, Ted Thomas apparently felt "there would have been no subsequent litigation." It was a pleasant thought.

Whether expressing his credibility finding in different words—presumably ones that would dilute its import—would truly have inoculated Justice Mahon against legal attack is doubtful. Gary Harrison, former counsel assisting the Royal Commission, always believed that since Mahon was a member of the establishment and had yet stood up to it, the establishment would have been "determined that he should pay for it in some way." Besides revealing an all-out national effort to cover up gross negligence on the part of the national carrier by aggressively pinning blame for the crash on the pilots, the Mahon Report had put the airline's reputation and the government's solvency on the line by retailing the particulars of that negligence. If his powerful enemies couldn't retaliate against the judge for telescoping months of untruthful testimony by the airline witnesses into a catchy five-word phrase, they were certainly angry enough to identify some other way to retaliate against him.

Still, "an orchestrated litany of lies" has confused some people, probably because it does not directly link two sets of fictions—the second one of which (litany of lies) was designed to reinforce the veracity of the first. That second fiction was the airline executives' claim (at the hearings) that the flight track for the past three years always went over Mount Erebus, for which reason pilots had to fly no lower than sixteen thousand feet until safely on its far side. The first fiction was Chief Inspector Chippindale's attribution (in his official report) of the crash's cause to pilot error because Captain Collins and first officer Cassin descended too low on *the normal* track over Erebus to avoid smashing into the mountain.

An associate of Justice Mahon later conjectured that he had never properly explained the "litany of lies" because he believed it self-evident that the airline executives would shortly be prosecuted for perjury. These men had lied under oath to him, after all. But Justice Mahon had not counted on Prime Minister Muldoon stealthily arranging for the police officer heading the probe merely to go through the motions—and ultimately determine that no perjury had taken place.

In any event, Mahon himself considered it imperative to put on the record that he had been told a fraudulent story. "I had to make it quite clear," he would explain later in a radio interview, "that I had been given [by the airline's executive witnesses] a very long and involved series of explanations about (1) altitude limits and (2) mistakes on the part of the navigation section, and that I did not accept considerable sections of the evidence given on those two points." Besides, a finding as to causation necessarily went hand in hand with a finding as to credibility; they were two sides of the same coin, especially in the Erebus case since Chippindale's report and Mahon's report were irreconcilable. This circumstance alone ensured that the resolution of the cause and culpability questions boiled down to who one believed: Chippindale or Mahon. In this context, the credibility of each party was essential to assess. As the judge understatedly put it, "I don't believe a Commission of Inquiry into an incident can perform its duties satisfactorily without the power to make findings about the credibility of the people who appear before it." By that standard, his arresting literary flourish in perfect iambic pentameter was downright inspired.

PART THREE

POLITICAL AND LEGAL FIREWORKS

13

SPARRING AFTER THE MAHON REPORT'S RELEASE

ROYAL COMMISSIONER PETER MAHON HAD JUST completed the country's highest level of public inquest into an unimaginable tragedy that had precipitously ended the lives of all on board a "routine" sightseeing flight to Antarctica. His official report's substantive findings and attribution of culpability were completely at odds with those of the government's in-house investigator, Chief Inspector Chippindale. Justice Mahon knew that his own independent analysis on behalf of the public interest would prove anathema to Air New Zealand's executives as well as the airline's lawyers, with all of whom, along with the head of Civil Aviation, he'd been contending throughout the Royal Commission hearings. The Mahon Report was also bound to set off fireworks in the Beehive, the informal name of the executive wing of New Zealand Parliament Buildings in Wellington.

As the hyperprotective nominal owner of the airline, Prime Minister Muldoon had badly needed from his appointee a second finding of pilot error if both Air New Zealand and its regulator were to avoid reputational damage and the very real possibility of legal liability and hence large insurance payouts. Instead, the royal commissioner had just done the utterly unimaginable. He had absolved Captain Collins and first officer Cassin of any responsibility for TE901's crash and, instead, had found Air New Zealand's appallingly slipshod organizational protocols and Civil Aviation's inadequate oversight of the airline responsible for one of the world's worst aviation disasters.

Infuriated that the man he'd selected as royal commissioner had failed to deliver the verdict expected of him, Prime Minister Muldoon launched his offensive as soon as the Mahon Report was released. "The findings do not accurately appear to be in accord with the evidence," he groused in an opening salvo. "In other words, the evidence does not lead to some of the conclusions that Justice Peter Mahon arrived at." His criticism prompted an Australian commentator to observe, "Prime Minister, you've attacked this Royal Commissioner, but we understand when he was appointed he was a distinguished judge." To this Muldoon snapped, "That is what we thought at the time." On another occasion the prime minister hotly declared, "We should have appointed more than one judge instead of just the one running off on his own."

Years later, former attorney general Jim McLay would thoughtfully look back on an explosive, bitter period without precedent in New Zealand history. On the one hand, he now wished multiple commissioners *had* been appointed. "I don't think two other commissioners would ever have put their signature to the 'orchestrated litany of lies.'" To the contrary, they would have "leavened some of

Mahon's language." On the other hand, he was stunned by Prime Minister Muldoon's rashness and presumption when the Mahon Report came out. "It was in my view if not inappropriate then certainly unusual for a prime minister to say, 'We appointed this commission of inquiry. It has reported A and I think not A.'"

Already at the time, astute observers had been quick to detect something amiss in Muldoon's fierce reaction to the Mahon Report. The opposition leader Bill Rowling, for instance, wondered whether the prime minister's objective was "looking after a clutch of personal political friends." He had "thought from day one that Rob Muldoon was far too quick to come to the aid of both Air New Zealand but [*sic*], more importantly, Morrie Davis." When asked at a press conference whether he might have left the impression that he supported Chief Inspector Chippindale's findings against those of Justice Mahon, the aroused Prime Minister Muldoon's ominous comment was, "What we are going to do is get this thing *tidy* one way or the other."

Made royal commissioner for the Erebus inquiry on Prime Minister Muldoon's own recommendation, Justice Mahon had now been abandoned by him. Intuiting that the man was itching for a public brawl, Mahon decided to give him one. Prime Minister Muldoon, who routinely abused and belittled political opponents, had hotly dismissed the Mahon Report's findings as "unsupported by any evidence set out in [it]." In a highly unconventional move for a member of the judiciary, Justice Mahon called a press conference.

"The Prime Minister's comments are interesting but unsound," he announced. "The evidence cannot possibly be understood without reference to thousands of pages of testimony and hundreds of exhibits." A reporter quickly noted that in his understanding "the PM has not read the evidence" but "trusts the opinions of those who have." "The Prime Minister's confidence in some of his advisers is admirable," Justice Mahon replied evenly. "I can only suggest that he may find in due course that it has been sadly misplaced."

In the short run, the judge proved highly adept at publicly defending his position against that of his formidable opponent. Still, concerned the fusillade of prime ministerial invective might wear down the royal commissioner, friends lent their support. One of them was Sian Elias, who later would become chief justice of New Zealand's Supreme Court (after the country ended appeals to the United Kingdom's Privy Council in 2003). Mahon reassured her by noting, "Once, while on patrol in Italy [during World War II], I was ambushed by a number of Germans who opened fire. Instead of diving for cover, I stood where I was and fired back." In his extended, very public exchange of blows with Prime Minister Muldoon, Justice Mahon followed the same protocol, standing his ground and arguing his case persuasively in the public arena. Compared to walking onto a minefield to retrieve a friend whose feet had just been blown off, as Mahon had also done during the war, it seems that taking on the country's bullying head of government appeared a trifle.

Being as forceful and abrasive a personality as Prime Minister Muldoon, Air New Zealand's CEO, Morrie Davis, also reacted testily when the Royal Commission's findings became public. Appearing before reporters an hour after the Mahon Report was released, he denounced its "totally indefensible" contents as an attack on his professional competence as well as his personal integrity. "I reject entirely," Davis fumed, "any allegations that my performance of duties, giving of evidence or relationship to the giving of evidence by others, was in any way inadequate or improper." Nonetheless, under pressure from Des Dalgety, Muldoon's personal attorney and by now deputy director of Air New Zealand's board, he quickly stepped down. His sudden retirement was not an admission of guilt, Davis explained, but an attempt to "remove a focus point from the current controversy" for the sake of accelerating the company's recovery. It was an ignominious end to a forty-year climb from Air New Zealand office boy to the top.

Known for his bellicosity and total lack of people skills, Morrie Davis was memorably described as a "man's man"—that is, someone disinclined to share his personal feelings. A member of the airline's cabin crew nonetheless got a rare glimpse into those feelings when he paid an impromptu visit to the Davis residence one evening following his former boss's resignation. The whole family was clearly suffering. Morrie was receiving hate mail and even death threats. His daughter was being bullied at her law school. Stress, sadness, and depression were taking a toll.

━━━ ◆ ━━━

As soon as he read the "orchestrated litany of lies" paragraph of the Mahon Report, Attorney General Jim McLay realized that the New Zealand government had a big problem on its hands. Straightaway, he had forwarded the report to the police with instructions that its chief superintendent, Brian Wilkinson, have specific Air New Zealand executives interrogated regarding the accuracy of their testimony during the Royal Commission hearings. He did not inform Muldoon of his action, fearing that the incensed prime minister might try to nix a police investigation.

Could charges of an organized conspiracy of deception and perjury by specific Air New Zealand employees be sustained? It remained to be seen. The airline had already placed twelve men who had testified at the Royal Commission hearings on restricted duty pending the outcome of chief superintendent Brian Wilkinson's probe. Among them were Messrs. Gemmell, Wilson, Johnson, and Hewitt.

The first person the police called in for questioning was chief navigator Brian Hewitt, who continued to protest that his shifting of the flight path to McMurdo Sound was a mistake, nothing else. If so, the witness was asked how he explained the fact that the "erroneous" flight path's destination waypoint coordinates had instantly turned up in new track and distance charts and in revised briefing instructions to all Antarctic pilots? Chief navigator Hewitt first suggested that it was a mere coincidence, then unburdened himself. "It seems incredible to me that [my error] should be used in a briefing document when it doesn't follow the correct route. Absolutely incredible. The whole thing is just amazing."

Captain Ross Johnson was second to be interviewed. Why, the police wanted to know, did he claim to have described a route over Erebus to five pilots in a briefing on November 9, when Captain Simpson had testified that he had described a route up McMurdo Sound? "I still think what I said was correct," Johnson answered. "Which would make Simpson's testimony incorrect?" "I might have been mistaken—or he might have been."

Despite Johnson's evasiveness, chief superintendent Wilkinson would conclude that this airline executive's negligence was responsible for the true magnitude of the fatal last change in the flight path going unnoticed. Captain Johnson was aware of Captain Simpson's reported *twenty-seven-mile* cross-track difference. He was also aware of Lawton's and Hewitt's miniscule *two-mile* cross-track difference. No alarm bell went off—he failed to notice the discrepancy.

The upshot of the police investigation was that the team's officers seemingly could not lay their hands on the level of corroborative evidence needed, in their view, for a charge of organized conspiracy of perjury to hold up. They came closest to action with chief navigator Hewitt and the airline's flight manager (line operations), Captain Johnson, against both of whom the charge of manslaughter was contemplated. Ironically, the two were reportedly spared through the inadvertent intervention of the royal commissioner. In the Mahon Report, he had famously declared that the accident was attributable not so much to any individuals but to deficiencies in the administrative and organizational arrangements of Air New Zealand itself. In the end, the police brought no criminal charges against anyone at the airline, and the twelve employees on hiatus returned to work.

That the police investigation had not been prosecuted vigorously was corroborated by the dispiriting experience of Sergeant Gilpin, who, along with Constable Leighton, had been among the first officers to arrive at the crash site back in 1979. As disaster victim identification (DVI) squad members during Operation Overdue's Antarctic phase, they had discovered Captain Collins's ring binder and its fully intact, highly revealing contents. Only in 1981 did Gilpin learn that the binder's pages had subsequently disappeared.

This intelligence had come to Sergeant Gilpin in the form of a documentary on the Erebus disaster. Incredibly, there on the TV screen was the royal commissioner, Justice Peter Mahon, waving a small ring binder with no pages inside as he interrogated an Air New Zealand witness, one Captain Bruce Crosbie, about where its contents had gone. As liaison officer to the Collins and Cassin families, Crosbie had been the one to return the ring binder, empty, to Maria two weeks after the accident.

> "How could the ring binder itself be intact yet the pad of writing paper disappear?" Mahon asked incredulously.

> "I suggest the cover survived the water and kerosene but the paper didn't," the witness volunteered. Of course Gilpin knew the man was lying!

In the video, Mahon tried again. "How do you suppose the pad of paper secured by the ring binder could have disappeared?"

"I have no idea," Crosbie replied. "As I say, unless they were removed because they were damaged. That would be the only reason."

Counsel assisting the commissioner now asked, "If papers were removed from the ring binder, who would have done that?"

"I would have myself, I presume," Crosbie declared. While having no specific recollection of doing so, the witness added, "I was involved in destroying a lot of papers that were damaged."

Watching the documentary, Sergeant Gilpin had realized that the airline's top executives must have grasped the significance of the ring binder's contents—that is, their potential to expose its own damning blunder with the undisclosed final waypoint's coordinates. They had apparently preferred to have Captain Crosbie tamper with evidence rather than allow the truth to come out. Gilpin tried to contact chief inspector Wilkinson's team, then investigating certain airline executives for possible coordinated perjury. Here was a welcome opportunity to share his and Constable Leighton's important information about the legibility of the contents of Captain Collins's ring binder at the time of its recovery. Strangely, the Wellington police officer was never able to connect with his colleagues in Auckland.

Deeply disturbed, Sergeant Gilpin later reached out to Captain Vette and Justice Mahon. There could be no doubt, the latter informed him, that the ring binder he'd seen in the documentary was in fact the notebook Constable Leighton had found in the vicinity of Captain Collins's body at the crash site. The reason its contents had been destroyed, Mahon suggested, was surely that they included the "navigation data which Captain Collins was given at the briefing for the fatal flight." A smoking gun, in other words.

The theft of the contents of Captain Collins's ring binder caused him "more heartache and stress than the body recovery operation," Sergeant Gilpin would confide years later in an interview, because he knew how wrong it was. He had made such a big deal out of the ring binder over the years because he "wanted to see justice for the crew and for those people who died." Sergeant Gilpin conceded that the ring binder's missing contents constituted only one small piece of a much larger puzzle. Nonetheless, he could vouch for Constable Leighton and himself, who between them eventually amassed over eighty years of professional experience. They knew what they saw, and they were distressed that it "was not dealt with correctly and someone interfered."

Only in 2012, shortly before he died, would Captain Gemmell, Chief Inspector Chippindale's technical adviser, admit that the airline had confiscated the ring binder's contents—but merely, he claimed, because they contained information regarding extramarital lovers. In an undated statement to airline CEO Morrie Davis, Captain Crosbie had long since explained that *he* had tossed out the ring binder's contents because they consisted simply of "references of a general nature such as shopping lists." For the record, Maria Collins would later state that her husband kept pertinent technical information in the ring binder. Praising Sergeant Gilpin for being "honest to the last part of his being," she slyly noted that "*he* knew the difference between a shopping list and coordinates."

By the time that the desultory results of the police inquiry came out, they were not unanticipated by Justice Mahon, who was an old friend of chief superintendent Brian Wilkinson. From him, the judge had learned that Prime Minister Muldoon was "counter-punching" the police investigation launched without his knowledge by instructing Wilkinson merely to go through the motions. He was to conduct his inquiry as called for by the attorney general but in the end make the case against the airline's suspended functionaries quietly disappear. To that end, the superintendent would have understood that his team's being in touch with Sergeant Gilpin was contraindicated.

Meanwhile, as the police investigation into specified Air New Zealand personnel ran its course, some were calling for Captain Kippenberger's dismissal as director of Civil Aviation. In response, the State Services Commission initiated an internal probe—*how* energetic is questionable—to determine whether any of the division's personnel should be charged under the State Services Act of 1962. Released in June 1981, the report found "significant short-falls in the overall performance of civil aviation in relation to the Antarctic flights." However, its authors concluded that no act of this division of the Ministry of Transport had made it in any way culpable for the Erebus tragedy. Kippenberger kept his job.

Under normal circumstances, a trenchant analysis of the cause of and culpability for a colossal air disaster—especially in a case as technically challenging and emotionally fraught as Erebus—would have been widely admired by Justice Mahon's fellow judges. However, the circumstances were not normal. Therefore, even as the judge was being elevated to the status of a folk hero by the New Zealand public at large, a fervid element within the judicial and political establishments was plotting to "kill the messenger" of a highly embarrassing message possessing portentous reputational and financial implications for the Muldoon administration. The campaign began but did not end with social ostracism. Justice Mahon gradually found himself abandoned by certain old friends and associates in high places who no longer dared (or cared) to associate with him. According to Margarita, all but two of her husband's colleagues on the High Court distanced themselves from him, dissolving into the judicial bench work.

An especially painful consequence of faithfully fulfilling his responsibilities, as Justice Mahon understood them, was the demise of his close friendship with high-profile Auckland attorney Lloyd Brown. It was he who had originally recommended Peter Mahon to the Muldoon administration as a potential royal commissioner for the unavoidable "independent" inquiry. After serving as the airline's lead attorney at the hearings before Justice Mahon, Brown must have been deeply chagrined that he'd been wrong about his friend's reliability. When, therefore, Mahon wrote him "suggesting lunch, after the dust had settled," Brown replied that it would "look bad with Air New Zealand" if he were to be seen with him. Gary Harrison, junior counsel assisting the Royal Commission, was in Justice Mahon's chambers for the deathblow. "Mahon replied that he was more concerned with the deaths of 257 people, and that was that." The two would never speak again.

14

THE DEPENDABILITY
OF BETRAYAL

AS A STUDENT OF SHAKESPEARE AND DANTE, AS well as a frontline infantryman during World War II, Mahon possessed a deep understanding of the human condition and, over the decades, conscientiously made allowances for a variety of revealed human weaknesses in his legal judgments. He seems to have entertained no illusions about mankind's worst tendencies and persistently to have extolled the law as society's mechanism for holding them in check. That a member of his own chosen temple might now—Iago-like or Brutus-like—be poised to betray him, the judge was keenly aware. He had a plan for that contingency.

So noxious were the Mahon Report's conclusions in the eyes of Air New Zealand's management that an unprecedented effort to delegitimize the Royal Commission, aided and abetted by certain of the judge's disaffected brethren in the senior ranks of the judiciary, now inexorably got underway. It signaled a transition away from the royal commissioner's quest to uncover the facts surrounding the Erebus tragedy and forthrightly report them to his compatriots. The transition was *toward* a brutal period of seemingly nonstop litigation to establish whether Chief Inspector Chippindale's official accident report or Justice Mahon's official accident report would prevail. The government's in-house inquiry had found that the national airline did not cause the colossal disaster in Antarctica—the pilots did. The royal commissioner's inquiry maintained the exact opposite.

Because Justice Mahon's was a fact-finding mission, not a criminal trial, Air New Zealand could not directly appeal his findings. The carrier's counterattack would thus of necessity take the form of attempting to catch him on a technicality. With the backing of Prime Minister Muldoon, the airline thus instructed counsel to initiate legal proceedings to dismiss the inquiry's findings in the circumscribed area of "staff integrity"—that is, Air New Zealand personnel's degree of truthfulness when testifying under oath at the Royal Commission hearings. This action, undertaken at the public's expense, was the closest the airline could get to a defamation suit against Justice Mahon.

The judicial review was originally scheduled to be heard in the High Court, which, despite the judge's status as one of its members, was normally considered the appropriate venue. Not everyone was satisfied with this arrangement, however, which prompted certain determined parties to invoke a special legal provision whose effect was to allow the case to be removed from the High Court into New Zealand's appellate court, the Court of Appeal. Justice Speight and others on the lower court protested to no avail. "The magnitude of the disaster," "the public importance of the issues," and "the conduct of an inquiry held by a High Court judge" were cited to justify this atypical legal course.

Did it matter who reviewed Air New Zealand's complaint against the Mahon Report's assessment of its executives' credibility?

The airline certainly thought so, which led some observers to ponder what its motives might be. Soon enough it would come out that, of the Court of Appeal's five justices, two were conflicted in having adult children who worked for the airline. Whether that would necessarily have affected their thinking about the Mahon case, we cannot know. Nonetheless, these members—Justices Woodhouse and McMullin—should have recused themselves under the "justice must be seen to be done" rule. In other words, if fair-minded people could conclude that there was a real likelihood of bias on these men's parts, they were obliged to disqualify themselves from judging the case.

Knowing about one of the conflicted jurists, Justice Mahon informally approached the Court of Appeal's acting president on behalf of the concerned NZALPA. No, Mahon was told, that judge insisted on sitting. NZALPA subsequently made a vigorous legal effort of its own to get this Court member removed from the case but was unsuccessful. At the time, Justice Mahon was not too worried: "he's [only] one out of five" who would conduct the Court of Appeal review of the Royal Commission's findings with respect to the airline witnesses' integrity. There were two out of five, though. In an interview with the *New Zealand Listener*, some months after the review had concluded, Justice Mahon would rue his ignorance of the second conflicted party, who was none other than the Court of Appeal's president, Sir Owen Woodhouse. Had he known, he said, "I'd have objected in person. Myself." Given what would prove to be Justice Woodhouse's outsized impact on the course of all future Erebus legal proceedings, Justice Mahon's misgivings were not unjustified.

Things did not get off to a good start when the Court of Appeal met in October 1981 for an unprecedented review of circumscribed findings in the Mahon Report. Right off the bat, the airline's lead counsel, Lloyd Brown, objected to the presence of Justice Mahon's counsel (the two individuals assisting him as royal commissioner). Justice Woodhouse was presiding, and, supporting Brown, he denied Mahon's counsel the right to enter and participate in the hearing. A huge argument ensued. It was resolved only when the solicitor general was sent for and supported their inclusion. "Without that, there would have been no effective opposition to Air New Zealand," erstwhile junior counsel Gary Harrison would later explain.

Mahon had earlier predicted that the attack against him would fall under the rules of "natural justice." This was a legal area in which he was a leading authority and whose principles his own decisions had helped to define. Mahon's intuition was correct.

The crux of the airline's complaint was that certain Air New Zealand personnel testifying before the Royal Commission were later depicted in the Mahon Report as having engaged in "a predetermined plan of deception" and "an orchestrated litany of lies." The company's contention was that these witnesses should have been (a) informed of the judge's adverse provisional credibility finding during the hearings and (b) given a chance to respond to it. Affected individuals included the airline's CEO, Davis, of course, but also Captain Eden, director of flight operations, and Captains Gemmell, Grundy, Hawkins, and Johnson—all of whom were executive pilots. The four members of the Navigation Section of Flight Operations (Amies, Brown, Hewitt, and Lawton) were also included.

Justice Mahon had crafted his credibility finding and delivered it publicly *subsequent to* the hearings after weighing several factors. The first of these was his determination to deny his opponents a pretext to charge him with bias and attempt his removal as head of the Royal Commission. Even as he was conducting his own investigations into the Erebus tragedy, another sitting royal commissioner in Auckland was running into headwinds in his murder inquiry for just this reason. Being from New South Wales in Australia, Justice Taylor was accustomed to a lusty legal tradition that allowed a judge to call out witnesses he believed were lying. Operating in that tradition, Justice Taylor had accused the police witnesses at his hearings of attempting to deceive him. In Auckland this action had the effect of galvanizing the police's lawyers to apply to the High Court to stop the proceedings on the grounds of bias.

Mahon had not been about to fall into the trap of accusing the airline witnesses of perjury when not all the evidence was in yet. He was at pains to convey to those who thought his mind was made up against these witnesses that he had, in fact, reached no firm conclusions. Nonetheless, Justice Mahon did use two back channels—David Williams, a counsel to the airline, and Peter Martin, a representative of Lloyds of London, the company's insurer—to warn Air New Zealand that the credibility of those testifying on its behalf was in serious doubt. The Court of Appeal presumably was unaware of the cautions Mahon discreetly had given behind the scenes. The ironic upshot was that, as Erebus authority Stuart Macfarlane has observed, "Justice Mahon's scrupulous attempts to be fair to the parties and witnesses before him by steadfastly avoiding any semblance of bias enabled the Court of Appeal to hold that he had breached the rules of natural justice through not making clear to the Air New Zealand witnesses that he disbelieved them."

A second factor that influenced Mahon's timing for the disclosure of his credibility finding was his understanding of natural justice's requirements. These, in his professional view, did not entail any obligation on his part to alert witnesses to a contemplated adverse finding, much less a duty to allow them to respond in additional testimony. There was a case brought in England that Mahon considered equivalent to that of Erebus in its dynamics. In that inquiry—like Mahon's, it was not a trial and hence had no accused—the investigating inspectors were tasked with discovering what had happened and reporting their opinion with, as Lord Denning put it, "courage and frankness." The sticky part was that this English case, like Mahon's, involved someone who was going to come in for censure. In the interests of fairness, was anything owed that individual before the adverse finding against him became public knowledge?

A fair inquiry, it had been suggested, contained three parts. First was hearing evidence and examining documents. Second was reaching a tentative conclusion. Third was putting the gist of the conclusion to the witness. The fascination of the case cited by Mahon is that it held that natural justice did not demand all three parts. In particular, it didn't demand the third part. Lord Denning explained that including witness rebuttals to tentative adverse findings against them would only produce a distracting sideshow. The proceedings would swiftly devolve "into a series of minor trials," prolonging the inquiry indefinitely. "I do not think it is necessary," Denning concluded. "It is sufficient for the [investigating] Inspectors to put the points to the witnesses as and when they come in the first place. After hearing the evidence, the inspectors have to come to their conclusions . . . [which] can be final and definite, ready for their report."

In a similar vein in the same English case, Lord Justice Lawton declared that he had not discovered "any other case which has suggested that at the end of an inquiry those likely to be criticized in a report should be given an opportunity of refuting the tentative conclusions of whoever is making it." In his view, those who conduct inquiries "are no more bound to tell a witness likely to be criticized in their report what they have in mind to say about him than a judge sitting alone who has to decide which of two conflicting witnesses is telling the truth." Inquirers naturally have the duty of giving witnesses with credibility issues an opportunity to correct misimpressions during the inquiry. However, "I see no reason why they should do any more." Justice Mahon had obviously agreed with this approach, for it is the one he deployed in his capacity as royal commissioner probing the Erebus disaster.

As it happened, however, the winds of change on a "fine point of law" were blowing into New Zealand from England at just this time. John Burn, Peter Mahon's close Christchurch friend, has described an academic movement afoot to impose on local judicial inquiries both the *duty to disclose* at the hearings themselves that certain witnesses were not being found credible and the *duty to grant* these parties an opportunity to testify again to try to change the presiding judge's mind. Although Justice Mahon had had no use for these notions when writing his own report, they would loom large in the Court of Appeal's review of it.

A third factor Mahon weighed in timing the release of his credibility finding was the danger posed by any act of his that would allow individual witnesses to prolong his inquiry indefinitely via a distracting legal sideshow whose only achievements could be further to muddy the waters and exhaust the public's interest in the Erebus case. This was a man on a mission. He wished to get out, fast and unambiguously, his discoveries that (a) Air New Zealand's unsafe management and operational practices had caused TE901's aerial misadventure and (b) a concerted disinformation campaign was afoot whose goal was to conceal the airline's culpability for the tragedy by misrepresenting it as a case of pilot error.

In short, Justice Mahon believed that his understanding of the nature of a royal commission and the requirements of natural justice in that context were correct. Four decades later, Ted Thomas (by then retired Court of Appeal judge Sir Edmund Thomas) would recall that his friend and colleague "was completely convinced, one, you couldn't review a Commission of Inquiry finding—that it was beyond judicial review—and two, that [the Court of Appeal] would substantiate his finding" by agreeing it wasn't out of order for him to say Air New Zealand had been guilty of concerted perjury.

Whether the appellate court's five judges were indeed prepared to endorse his views on these two matters, Justice Mahon was about to discover.

———— ◆ ————

Hamilton district court judge Peter Spiller is unusual in being a former university professor who holds six degrees and two doctorates. In his academic career, he was known to be an outstanding scholar, teacher, and mentor. A top legal authority on the Court of Appeal and the Privy Council, Judge Spiller has interviewed many members of both bodies, read numerous decisions and judgments that have gone on appeal, and authored an authoritative history of the Court of Appeal in the period 1958–1996.

During his extensive research for that impressive tome, Judge Spiller reports having discovered an interesting phenomenon: "The language and tone used by the original decision-maker can play a role in how the decision/judgment is viewed by the higher court." More specifically, "If the language used is restrained and gives the appearance of balance and objectivity, the decision/judgment is harder to undermine." In the Mahon Report, by contrast, the royal commissioner had deployed intemperate language in the "stance of the airline" section, which had concluded with a sensational finding that certain airline witnesses had engaged in coordinated perjury at the hearings. Here was an opening for any Court of Appeal member so inclined to respond to Mahon's "language and tone" in a similarly incendiary manner by attacking and attempting to discredit *Justice Mahon's* credibility.

Between the High Court's Justice Mahon and the Court of Appeal's president, Sir Owen Woodhouse, there subsisted a longstanding antipathy that superficially seemed odd in view of their shared status as returned World War II servicemen and distinguished senior judges known for their literary flair and witty bent. Despite these commonalities, they differed greatly in judicial outlook, with Mahon making no secret of the fact that he objected to the strain of judicial activism characteristic of his Court of Appeal colleague's judgments. According to Sam Mahon, who read hundreds of his father's personal letters from the late 1970s and into the 1980s, a recurring theme was "a creeping dislike of Woodhouse, somewhat similar to rust." Over time, his son ominously wrote, there evolved between the two a "constant undermining of each other" that seemed to be building toward some sort of crisis.

Both Justice Mahon and Justice Woodhouse had "strong personalities" and "could be blunt in expression," Judge Spiller has observed. Woodhouse was, in his professional view, "the most liberal judge of the [Court of Appeal] to become President." Furthermore, "This quality was reflected in his flexible approach to legal authority and his preparedness to extend the law to meet the demands of practical justice and societal needs." *Extend* the law?

However the rivalry and veiled personal animosity between the two luminaries had arisen, it was to prove highly consequential once the airline's legal action was removed from the High Court into the Court of Appeal. For his part, the erstwhile royal commissioner suspected that Woodhouse had "made up his mind long before the hearing that he was going to attack my report." Portentously, Mahon confided to a friend that "if Woodhouse announced there were no grounds for my credibility findings . . . I would take that as a vote of no confidence and hand in my portfolio." Such a determination would, he believed, impugn his very capacity to judge—to ascertain who was lying and who was telling the truth. A potential crisis thus seemed to be brewing, for Mahon, who was said to possess "a seer's ability to divine the thoughts and motivation of witnesses," took justifiable pride in his courtroom skills as a detector of falsehoods.

As president of the Court of Appeal, Justice Woodhouse was entitled to draft its judgment, and in due course he produced around 160 pages scathingly critical of the Mahon Report. An exceptionally strong statement was necessary, so he would later tell Spiller in an interview, to protect "innocent people"—presumably the top personnel at Air New Zealand—affected by Mahon's "obsessive" charges. Although the Court of Appeal's was strictly a *review* case, Justice Woodhouse in his draft judgment ventured to "extend" the law by critiquing the merits of Justice Mahon's factual findings as well.

Justice Woodhouse's draft proved unpalatable to the rest of the appellate court's members. Three of the judges (Cooke, Richardson, and Sommers) believed that besides its uncomfortably strident tone, the draft inappropriately addressed the Royal Commission's factual findings, which were off limits. The trio thus decided to write their own judgment, one separate from that of Woodhouse. As for the remaining member of the Court of Appeal, Justice McMullin, he had ingratiated himself with its president and was prepared to join him in a minority judgment—but only after the latter had agreed to delete parts of his document and soften its tone. According to Judge Spiller, "The case caused unfortunate tensions and strains in the court, and the period around the time of the judgment was unpleasant."

While the appellate court thus rendered two judgments, its members were unanimous in holding that the royal commissioner's implied powers were not so comprehensive as to entitle him, in his official report, to charge senior airline executives with having engaged in a conspiracy to lie during their testimony before him. The judges were likewise unanimous in ruling that Mahon had breached natural justice by giving airline witnesses of dubious credibility no opportunity, during the hearings, to defend themselves against his provisional adverse findings, of which they had not been made aware. The upshot was a third unanimous Court of Appeal decision: to quash the royal commissioner's costs order. Ordinary citizens would understand the fine, if allowed to stand, as a punishment imposed on Air New Zealand for needlessly prolonging the inquiry by instructing its executives to tell Justice Mahon lies. In the view of all five members, the reputations of those adversely affected by Mahon's "orchestrated litany of lies" finding would be vindicated by their having set aside the costs order.

According to recollections Margarita Mahon would share years later, her husband wasn't thrilled but could live with the Court of Appeal's majority opinion, known as the relatively benign Cooke Judgment. Judge Spiller has characterized it as "more balanced and dispassionate" than the court's fiery minority opinion, referred to as the Woodhouse Judgment. Alarmingly, however, the "bitter personal and professional blow" struck by the Woodhouse Judgment provoked in Justice Mahon a drastic reaction. Among other things, it plunged him into a public controversy with airline CEO Morrie Davis, during which row they both gave television and press interviews. According to Spiller, Justice Mahon was also in extensive communications with the Muldoon government.

In retirement after his own illustrious judicial career as an appellate court judge, Ted Thomas would explain his old friend's impassioned behavior after the Court of Appeal judgments came down as stemming from shattering disillusionment that "his work of a lifetime" had just been maliciously and vindictively attacked by two of the Court of Appeal's five members. Thomas's "favorite judge of all time" was Woodhouse. However, he took exception to his hero's actions in connection with the Court of Appeal review, beginning with his failure to disqualify himself from judging the case to avoid any semblance of personal interest or bias. It was singularly unfortunate, Thomas added, that Justice Woodhouse felt emboldened "to remake the facts" in the process. That tampering with the facts had grossly distorted the minority judgment's analysis of the accident's cause, leading Mahon at the time caustically to characterize it as "nothing more than a disguised approbation of the Chippindale Report," which merely reflected "the attitude of the management of Air New Zealand" itself.

15

THE WOODHOUSE EFFECT

WHAT DID JUSTICE MAHON MEAN TO CONVEY BY DIS-
missing the Woodhouse Judgment as a "disguised approbation of
the Chippindale Report?" He himself had nothing but contempt
for that report and had, as a matter of fact, already thoroughly
demolished its thesis of pilot error as the cause of TE901's fiery implosion on the flanks of an Antarctic volcano. What had infuriated
the judge the most about the testimony of certain Air New Zealand
executives at the lengthy Royal Commission hearings was precisely
that the "orchestrated litany of lies" these witnesses had told him
was clearly designed to reinforce the correctness of Chippindale's
analysis. Was it merely by chance that said analysis was wholly
acceptable to the airline's top management? Mahon thought not.

"Follow the money and the information," an adage counsels, to
ascertain in a particular case what is financially at stake and who
at the top is calling the shots. Air New Zealand was a state-owned
company, which meant the entire government would be put at reputational and financial risk if the airline's hitherto unrecognized

management and operational dysfunctions were revealed and widely believed to be responsible for a colossal air disaster. All would be well for the Muldoon administration if something other than the airline itself could convincingly be blamed for the massive loss of life in Antarctica. The "pilot error" thesis expounded in the Chippindale Report perfectly fit the bill.

The Court of Appeal's president was determined to underscore the veracity of the Chippindale Report's analysis of the accident's cause by undermining the credibility of the Mahon Report's shocking "imputations of collective bad faith which had started from a high place in the company." He acknowledged that Air New Zealand's top officers had been legally barred from appealing Justice Mahon's findings on the accident's cause (they "must and do stand"). Nonetheless, after providing a detailed legal rationale for what he was about to do, Woodhouse proceeded energetically to attack those very causal findings as false. Regrettably, he alleged, they had led the royal commissioner erroneously to conclude that dysfunctional airline procedures had been responsible for the accident rather than pilot error.

In the court's minority judgment, Woodhouse denounced Mahon as a fabulist—a storyteller prone to misrepresenting the facts in support of utterly senseless conclusions. To prove the truth of this devastating charge, Woodhouse adduced illustrative examples of Mahon's supposed manipulation of factual evidence. The exercise was unfortunate because, in undertaking it, he twisted some verifiable facts, overlooked others, and even concocted "alternative" facts.

In the Court's judgment that bears his name, Woodhouse assumed (or pretended to assume) that Chief Inspector Ron Chippindale was an impartial investigator into the circumstances surrounding a horrific aviation disaster. He made no mention of the chief inspector's status as a governmental functionary suddenly caught up, along with other governmental functionaries, in a race against time to explain (or explain away) how a jet belonging to "nobody does it better" than Air New Zealand could suddenly kill a huge planeload of tourists. Therefore, to rebut a well-known allegation that Captain Gemmell absconded from the crash site with a blue envelope stuffed with documents, Woodhouse claimed that since Chippindale never complained about such a missing file, the charge must have been spurious. If evidence Gemmell was told to collect for him had never, in fact, reached the chief inspector, Woodhouse maintained that he "would seem to be the first person who would want to know how and why." Besides, he added, Gemmell had always been roped to others as he moved around the crash site, giving him no opportunity to steal evidence. The Court's president conveniently neglected to mention that potentially critical documents recovered at the crash site were transported for safekeeping to a special storehouse nearby. As Chippindale's technical adviser, Captain Gemmell could have accessed it at any time.

According to Woodhouse, Chippindale did perceive some dysfunction at Air New Zealand and forthrightly incorporated it into his official report. Most significant was the chief inspector's concession that materials issued at Captain Collins and first officer Cassin's route qualification briefing could have been misleading, inducing them to believe that the flight track was up flat McMurdo Sound instead of over a towering Ross Island volcano. However, just before the accident aircraft departed, Flight Dispatch corrected any possibly mistaken understanding when it gave the pilots a mass of updated flight path data to type into the aircraft's onboard computer. Woodhouse made no mention of the fact that, in contravention of international flight safety protocols, the pilots were not informed of the twenty-seven-mile longitudinal alteration of the revised destination waypoint. (For technical reasons, they would have been unable to detect the change on their own while inputting the updated nav track's coordinates.) Instead, Woodhouse commended airline CEO Morrie Davis for swiftly turning over to the government's "well equipped," in-house chief inspector of air accidents the important task of probing the disaster and determining its cause.

Highly incriminating evidence that, had it not swiftly disappeared, would have dramatically exposed Air New Zealand's responsibility for TE901's crash was first officer Cassin's paperwork from the briefing nearly three weeks before the accident flight. The day after the disaster, Air New Zealand had learned that Cassin had mentioned to Flight Dispatch that he'd left his briefing notes at home by mistake. Was there anything suspicious in the subsequent behavior of the liaison officer quickly sent to the Cassin home ostensibly to help the first officer's widow? One might have thought so since his principal interactions were not with Anne Cassin at all but with her in-laws. With their approbation—but without Anne's consent or even knowledge—he had taken away a mass of potentially revelatory airline paperwork.

No, averred Woodhouse, there was nothing at all questionable in Captain Crosbie's behavior. First off, despite Anne Cassin's allegations, Crosbie himself "denied ever receiving the [briefing] material" from the Cassin home. Even if he had, though, it would have been "irrelevant" in Woodhouse's sly reconstruction of events. Why? Because the Mahon Report raised no concerns about the line pilots' conduct. This was disingenuous because Woodhouse appreciated that the royal commissioner's sole interest in Cassin's briefing materials and notes concerning them was as a smoking gun corroborating that it was the airline, not the pilots, who bore responsibility for losing TE901 and all those cocooned inside it.

Woodhouse took jabs at the Mahon Report's finding on Air New Zealand's reprogramming of the destination waypoint in mid-1978, which moved it from the south side of Mount Erebus out into McMurdo Sound, twenty-seven miles away. On balance, the royal commissioner had considered the relocation deliberate but refrained from so ruling (correctly, as it would turn out) because of a small piece of evidence that did not square with this hypothesis. Still, owing to his logician's mind, Mahon said that even if the relocation was made accidentally, it was obviously far superior from a safety viewpoint to overflying a volcano. In his revised view of the matter, therefore, he claimed that even if there *was* a programming error, airline personnel embraced that error immediately for its soundness and practicality. But if it was such a superior route, Woodhouse demanded to know, then why couldn't Mahon convincingly explain why the airline would suddenly shift the destination waypoint back to the south side of Ross Island's Mount Erebus in the wee hours of November 28?

As it happened, the author of the Mahon Report had disbelieved the Navigation Section's testimony that its members thought they were moving the destination waypoint ten minutes of longitude (two and one-tenth miles)—not *two degrees and ten minutes* of longitude (twenty-seven miles). That was a contrived story, the royal commissioner had thought, designed expressly to explain away why Captain Collins was never told his flight path had been altered (an internationally recognized no-no). Here, however, Mahon was wrong. It seems that despite his intimate knowledge of how impaired the operations of Air New Zealand were, even his own supremely logical mind had limits on how much administrative and organizational dysfunction he could reasonably believe Air New Zealand had.

Woodhouse challenged Mahon's finding that the single change to the flight plan sent to US air traffic control on the morning of the accident flight was deliberate. (To reiterate, that change involved labeling the destination waypoint as "McMurdo" rather than providing its exact coordinates.) In the Woodhouse Judgment, the

Court's president claimed that neither Air New Zealand in general, nor the Navigation Section in particular, would have had reason to think Mac Center wouldn't approve a flight path over Mount Erebus. Here again, Woodhouse was prevaricating. For safety reasons, the Americans required all three military services (those of New Zealand, the United States, and Australia) operating in the air space they controlled to fly up the center of McMurdo Sound on approach to McMurdo Station and Scott Base. This being the case, it was inconceivable that Mac Center would ever countenance a *civilian* jet overflying Mount Erebus.

Despite Woodhouse's mischaracterization of the significance Mac Center would have read into a flight plan that explicitly showed a track over Mount Erebus, he was on firmer ground when he attacked Mahon's views on why the Navigation Section would purposefully try to fudge the fact that it was switching the track back over the mountain. Justice Mahon was never at a loss for possible explanations for questionable moves on the part of Air New Zealand. What he had suggested as the reason for the sudden appearance of a word where coordinates were supposed to have been was, though, involved enough to elicit Woodhouse's censure.

The truth of the substitution may have lain closer to hand than Justice Mahon imagined. The change may have been deliberate, of course, in which case it had deprived Mac Center of the opportunity to alert the accident flight's pilots that they were in mortal danger if they stayed locked onto their computerized navigation track. But what if the change had been accidental, just another Navigation Section operative's error in a long, "unorchestrated" litany of airline errors? Given Air New Zealand's numerous systems issues at this time, was that really too far-fetched an explanation for Mahon to have taken seriously? Possibly, he had.

A final matter to mention here is Woodhouse's charitable interpretation of the significance of Air New Zealand's failure to educate those commanding Antarctic charters about the perils of whiteout, which provided the proximate cause of TE901's crash into Mount Erebus. Describing the phenomenon as a "freak" rather than a common meteorological condition in polar regions, Woodhouse asserted that the airline did not deliberately ignore it "as a factor that should be taken into account in favor of the aircrew." Headquarters was merely "ignorant of the deceptive dangers of the whiteout phenomenon." This sleight of hand falsified the facts. Air New Zealand had plenty of intelligence on forms of whiteout—the company just did not share it with those needing it most. Besides, since when is being ignorant of deceptive dangers forgivable if a national airline is sponsoring tourist trips to a hostile, whiteout-prone part of the world?

———————

Once he had read the Court of Appeal's minority judgment, which reflected the views of Justices Woodhouse and McMullin, the erstwhile royal commissioner recognized the determined effort to discredit his evidence in support of an organized conspiracy for what it truly was: a way to declare his assessment not merely outside the ambit of his authority but also substantively invalid. Mahon took no solace in noting that virtually all the Woodhouse Judgment's evidential conclusions were wrong. What else, he confided to a correspondent, could one expect from the head of an august tribunal who had publicly expressed incredulity that the DC-10's crew could not actually *see* the mountain before colliding with it?

   Concerned friends tried to cool the aggrieved Mahon down before he could take some intemperate action he might later regret. His compelling analysis of the causes of the Erebus disaster was not being challenged by three out of the five members of the Court. Just two of them were behaving badly, these friends reminded him, and everyone knew why—they had intimate family ties with Air New Zealand through their children. Justice Woodhouse had a daughter who worked in the airline's public relations department, and Justice McMullin had a son who flew for the airline.

Besides the two men's underlying familial conflicts of interest, Justice McMullin had become entangled in another, more consequential one involving Air New Zealand's legal counsel. Assisting Justice Mahon's old friend Lloyd Brown, now the carrier's lead attorney, was a junior lawyer that was a qualified pilot. Serving as third counsel for the airline during the Royal Commission inquiry, he had the specific duty of organizing technical evidence to be presented at the hearings. In other words, his job was to brief airline executives on the very "evidence" that Justice Mahon would find unbelievable. This fellow, incredibly, was a close friend of Justice McMullin's family. According to Mahon's intelligence, he "was virtually living at the McMullin residence both before and during the inquiry."

Justice Mahon's friends' well-intentioned counsel fell on deaf ears. "He told me that he was going to resign," Ted Thomas would sadly recall. "I tried to talk him out of it, of course, and I got others to try and talk him out of it. But he was quite bent on resigning."

Without consulting his wife, Justice Mahon now took a step that would have serious financial repercussions for the couple: although there was no external pressure on him to do so, he submitted a letter of resignation to Attorney General McLay. (Since the judge's appointment had been by royal warrant, Mahon needed the attorney general's consent to leave the bench.) Despite uncomfortable pressure from Prime Minister Muldoon for him to consent immediately, McLay initially chose to ignore the letter.

For quite some time now, however, Mahon had been on an increasingly ugly collision course with Muldoon, who remained in high dudgeon that the royal commissioner had not reached the conclusions he expected his appointee to reach. The undignified exchanges between the two, which were occurring in the public domain, eventually led Attorney General McLay to reconsider his options with respect to Mahon. Close colleague Ted Thomas also found himself forced to reconsider his support for his friend "when he became vitriolic in his criticism of the Court of Appeal." Some thought that the judge's vigorous efforts to defend himself in the face of unprecedented lying were understandable, indeed justifiable. In other circles, however, it appears that he was now beginning to be perceived unfavorably.

Several decades later, Ted Thomas would recall his conflicted feelings at that fraught time. To him, Mahon was rather like a dog with a bone—"nothing could divert him." The more press coverage he got, moreover, the more impossible the situation became, "and he was showing no signs of letting up." Thomas, who had been conferring with Attorney General McLay about the matter, now regretfully "suggested that perhaps the time had come to accept Peter's resignation." For Thomas "this was a very tearful moment."

Justice Mahon himself bluntly explained to an empathetic correspondent that the Woodhouse Judgment's insufferable assessment of his judicial abilities had "decided me to resign." He added that with this act "I have made my effective protest." Then in a long, revelatory national radio interview with Sharon Crosbie broadcast on the first day of February 1982, Mahon reiterated that resigning in a blaze of publicity had been his response to "the minority thing"—that is, the Woodhouse Judgment. That judgment's thrust had been to endorse the airline applicants' contention that *no evidence* existed that could support a finding of a large-scale conspiracy to commit perjury. Mahon was apoplectic. He told Sharon Crosbie in their interview that "I don't believe in New Zealand's legal history that any . . . High Court judge has ever been told by a Court of Appeal that he has no grounds for a finding of credibility."

The former royal commissioner's tone of voice had moved from neutral through angry to outraged as he spoke.

In a later study of the Court of Appeal, Judge Spiller would describe *Re Erebus Royal Commission; Air New Zealand Ltd v Mahon* as "perhaps the most emotionally fraught appeal heard by the court in its history." Woodhouse had attempted to bar Mahon's counsel from attending the hearings but was overruled by the solicitor general. That the Court of Appeal could not produce a single combined judgment highlights just how contentious the hearing was—and just how prominent a role Justice Woodhouse played in making it so. "No grounds," Mahon said again to Sharon Crosbie about the minority judgment. "That was the finding that made me decide. It was the final straw."

"There is no hard and fast rule that a Royal Commission of Inquiry can *never* make a finding of organized perjury," Judge Spiller has opined, "if its head thought that that had occurred and [it] was clearly established." Woodhouse and McMullin, however, managed with their verbal pyrotechnics to cast doubt on the reliability of the royal commissioner's evidence in support of his credibility finding. It proved enough to persuade a swath of the New Zealand public, as well as some members of its legal and judicial fraternities, that a well-organized cover-up of Air New Zealand's responsibility for a massive loss of life in Antarctica might just have been a figment of Justice Mahon's imagination.

According to Justice Mahon's close friend Ted Thomas, then serving as president of the Auckland Law Society, the profession was shocked and dismayed that Mahon would abruptly resign in what appeared to be a moment of foolish impetuosity. He had been, everyone knew, one of the most distinguished and exemplary of High Court judges. The country would have benefited greatly by his continued leadership in a variety of areas. Sian Elias, future chief justice of the Supreme Court, wrote to Mahon on January 28,

1982, that she shared his "disgust at a system which would mete out such treatment"—and so discourteously too. The brilliant judge's untimely resignation in a retaliatory strike against the Woodhouse Judgment she considered a personal tragedy for her as well as an incalculable loss for the bench.

As passions became inflamed around the country, Peter and Margarita Mahon retreated to a quiet corner of Christchurch, where the judge courageously and defiantly wrote *Verdict on Erebus*. According to son Sam, that remarkable work was completed in a white heat in a mere several months, so intense was its author's drive and concentration. Here again Justice Mahon was breaking ranks, for the legal profession frowned on literary exercises involving a controversial career matter. For its author, though, the book afforded a golden opportunity to rebut the Court of Appeal's not altogether reliable findings by getting Mahon's own viewpoint and the facts informing it compellingly across to the public. Although, as a private citizen, he was no longer protected from being sued for libel, the judge dared to repeat in this popular account of his accident investigation many of the things he'd said in his official report. No one challenged their veracity.

16

APPEAL TO JUSTICE'S FOUNTAINHEAD

JUSTICE MAHON STRONGLY DISAGREED WITH THE Court of Appeal's determination that a Royal Commission of Inquiry had no right to include in its official report that some of the evidence amounted to organized perjury. It thus wasn't entirely surprising when rumors began circulating that Mahon might appeal that determination to the Privy Council in London. As Judge Peter Spiller, who is an expert on the Privy Council, has noted, the institution had long been "romanticized as the ultimate fount of justice." The basis of appeal to its Judicial Committee lay in the royal prerogative, "which allowed litigants in the last resort to pursue their grievances from their Sovereign's courts anywhere in the world" to justice's fountainhead in London.

In his radio interview with Sharon Crosbie early in 1982, Mahon had sounded eager to approach the Privy Council for the purpose of having his understanding of the nature of a Royal

Commission confirmed. From the broadcast, it appeared that he'd reached the momentous decision mainly on his own. When his old friend John Burn questioned whether an appeal was politic, he met resistance. "There was always a shell," Burn noted. "The long face, like a mask. There was a part of him I could never reach."

Based on his own analysis of Justice Mahon's prospects at the Privy Council, former attorney and Auckland law lecturer Stuart Macfarlane also believed the judge's contemplated course of action was unwise. He fared no better than John Burn, though, when he tried to warn Justice Mahon of the perils of making an appeal. In an article titled "Little Value Seen in Privy Council Appeal" that was published by *The Press* in Christchurch (where the Mahons were then residing), Macfarlane presciently explained why it was "inevitable" that the Privy Council would uphold the decision of the Court of Appeal. Moreover, he suggested, "if the government should encourage the Commissioner to appeal, say by offering to pay his costs, this would give the appearance of extreme generosity, while really having the effect of being a bait to the Commissioner to step into a trap."

Like most of his compatriots, Macfarlane had initially assumed Chief Inspector Chippindale's verdict of pilot error was correct. Later, he read the conflicting report of Royal Commissioner Mahon but, having no great interest in the topic, didn't pursue it. Only when the Court of Appeal's findings were made public at the end of 1981 was the incipient Erebus researcher jolted into a reassessment of his earlier views.

In reviewing the Court's minority opinion, known as the Woodhouse Judgment, Macfarlane was startled to read that Justices Woodhouse and McMullin maintained that Chief Inspector Chippindale had known perfectly well that TE901's flight path had been reprogrammed from McMurdo Sound to Erebus shortly before takeoff without notifying the aircrew—and "explicitly" had said so in paragraph 1.17.7 of his report. Macfarlane found, however, as had Justice Mahon before him in paragraph 48 of his own report, that Chief Inspector Chippendale's remarks concerning the last-minute coordinate change and the airline's attendant failure to inform the aircrew—specifically paragraphs 1.17.7 and 2.5—were by no means transparent. Indeed, they were in the law lecturer's estimation "a masterpiece of obscurity" that prevented the truth of the accident's dominant cause from straightforwardly outing.

It was this disturbing discovery that determined Macfarlane to make the Erebus case an extended research project, one that would turn him into the world's top authority on the air disaster and its incendiary sequelae. So intense was his focus on producing his massive tome of 736 pages, *The Erebus Papers*, that his wife said he did not come home to dinner for eight years. Published in 1991, this monumental work became the Erebus researcher's bible, providing both a handy compendium of valuable source materials and a critical assessment of the merits and demerits of each source's claims. On a philosophical or theoretical level, the work concerns nothing less than "the operation in practice of the system of justice" as well as "how that system deals with a major public disaster."

The fact is that Chief Inspector Chippendale so effectively disguised the real cause of the accident in his report that the public was not even aware he *had* disclosed it, albeit in maddeningly obscure language. Essentially what Chippindale asserted was that Captain White, who commanded the sightseeing flight of the week prior to Captain Collins's flight, was supposed to have been told about the planned route change but was not. Captain Collins was also supposed to have been told about the route change, now live in Air New Zealand's ground computer, but was not. In other words, Captain Collins was not told the same thing Captain White had not been told the week before.

The sleight of hand here is Chippendale's creation of a false equivalency between the two flights. Yes, both pilots had been briefed on a flight path up McMurdo Sound and shown the same latitudinal and longitudinal coordinates for its destination way-point. However, Captain White's nav track was still taking him up pancake-flat McMurdo Sound, while Captain Collins's was now taking him over Ross Island on a beeline for the mountain. Not telling Collins what wasn't told White was therefore potentially consequential—catastrophic, as it would turn out, since low flying on approach to Hut Point Peninsula's scenic attractions had long been sanctioned by Air New Zealand and regulated by American air traffic controllers at McMurdo Station.

Macfarlane's advice to Justice Mahon to steer clear of the Privy Council fell on deaf ears. The latter was convinced both that the Court of Appeal had misunderstood the requirements of fairness with respect to a Royal Commission of Inquiry and that his own approach was supported "by both justice and law." About the trap law lecturer Macfarlane had warned Justice Mahon? He stepped into it.

In due course, the Privy Council's Law Lords did agree to hear Justice Mahon's appeal. Before the hearings were scheduled to commence, though, a portentous development took place, reported here publicly for the first time. Two members of New Zealand's Court of Appeal quietly called on the Law Lords in London. One was President Woodhouse, who had been a member of the Privy Council as of 1974 and had sat in the Judicial Committee as recently as May–August 1981. The other was his close colleague Justice McMullin, a more recent member of the Privy Council. Separately they flew the ten thousand miles to London to confer privately with the Law Lords.

If the pair's trips had become common knowledge, it would have been a bombshell: two members of the Court of Appeal with intense antipathy toward Justice Mahon had flown to London to present their view of how the Law Lords should rule on their antagonist's appeal to the Privy Council. Those same Law Lords later would, in fact, produce a report so bizarre and self-contradictory that it defied explanation until now. However, it just may be that the Privy Council's findings were very much in line with the private recommendations of Woodhouse and McMullin on their respective unpublicized visits.

No one knows what the Court of Appeal justices *said* to the Law Lords, of course. In and of themselves, visits to London by Court of Appeal justices were not unusual. Justice Mahon lodged no complaint about these trips (he did know about them) because, presumably, he had no tangible proof that his enemies had worked against his interests while in London. Still, it is probable that the shared objective of Justices Woodhouse and McMullin was to persuade the Law Lords that it was in the national interest for the Court of Appeal's verdict(s) to stand. Less controversially, under the "justice must be seen to be done" metric, these visitors had behaved badly indeed.

By contrast, Chief Inspector Chippindale's prehearing maneuvering ignited a brouhaha precisely because there *was* tangible evidence in the form of a letter. Chippindale had written the Law Lords on how best to present certain information to their lordships. When Justice Mahon learned of this missive, whose contents he considered defamatory, he and Attorney General McLay had quite a row.

In mid-1983, nearly four years after the Erebus tragedy, the Privy Council's Law Lords under the chair of Lord Diplock took up the emotionally charged Erebus episode in a packed London courtroom. Despite his initial hesitance to get involved in a case that raised questions of fact, Lord Diplock was persuaded to issue a full judgment, so Judge Spiller has explained, in the interests of producing "a distant and authoritative pronouncement that was seen to be fair to all parties." Reportedly he never worked so hard on a judgment in his entire career.

The Privy Council's Judicial Committee had, in fact, given itself a daunting challenge: how to find the pilots blameless (as Mahon had convincingly proved that they were) without finding the outraged management of Air New Zealand (of which the outraged prime minister was the sole shareholder) and its outraged regulator blameworthy in their stead. The Law Lords' ingenious solution was to embrace the royal commissioner's views on causation but not his views on the Air New Zealand witnesses' credibility. In other words, they accepted his factual verdict but not, paradoxically, the body of evidence on which it was based (since it implicated the airline in a cover-up of its culpability).

The upshot proved to be a curiously convoluted report that, among other deficiencies, in places defied logic (or deliberately ignored it). For example, had the briefing officers testified truthfully about the route always being over Mount Erebus, then the pilots were blameworthy for flying straight into it at low altitude. Had those officers lied at the Royal Commission hearings on this critical subject, however, then the pilots were not responsible for the crash. The Privy Council wanted it both ways. The flight path *was* over McMurdo Sound and Captain Collins *had* plotted his route up it the evening before the fatal flight (presumably because the briefing officers had indicated that the sound was the route). At the same time, however, the briefing officers had not participated, during the Royal Commission hearings, in an organized deception designed to suggest otherwise for the purpose of providing false support for Chief Inspector Chippindale's verdict of pilot error. But if at those protracted hearings the briefing officers were not deliberately lying about what they said on November 9, why did the three surviving pilots who attended that same briefing testify that they were? The Privy Council did not care to explain this inconsistency—and numerous others as well.

A second instance of weak reasoning occurred when the Law Lords attempted to sidestep Captain Crosbie's motive for repeatedly visiting the Cassin residence in the crash's immediate aftermath. From an employee at Flight Dispatch, the airline had swiftly learned that first officer Cassin's folder from the November 9 briefing session must be in his house. Along with any notes or memoranda associated with its contents, that folder constituted key evidence since it would show that the pilots had been briefed on a route up McMurdo Sound.

During his visits to the Cassin residence, Captain Crosbie removed a trove of the dead first officer's papers, including the November 9 briefing documents along with his ancillary, handwritten notes. In the Privy Council's disingenuous reconstruction of the event, Captain Crosbie obtained Cassin's documents from the *widow herself*. In fact, though, Anne had nothing to do with the transfer of documents into the liaison officer's hands. That officer acquired them through the good offices of her brother-in-law, who appeared much more willing than she to facilitate the removal of her husband's belongings from her house. Furious, Anne would make a very public fuss about this episode for years. She was last publicly heard from on the subject over three decades after TE901's crash when interviewed by Christine Negroni, an American air disaster journalist, for her fine book, *The Crash Detectives*. "Insurance details, Greg's time sheets, bills, receipts, personal letters, bank statements, check stubs, flying books—everything had gone."

While conceding that Air New Zealand had a right to certain of her late spouse's paperwork, Anne Cassin had been adamant that three pages of handwritten notes in his briefing folder were private property. Captain Crosbie claimed never to have seen them, yet they had disappeared along with the rest of her husband's briefing-related and other documents. If the liaison officer had not taken them, Anne wanted to know, where were they?

Regarding the speedy removal from the Cassin residence of documents Air New Zealand knew to be highly incriminating, the royal commissioner believed the widow and disbelieved the airline official. The Law Lords, however, chose to believe Captain Crosbie rather than Anne Cassin. Why? Without attempting to square their conclusion with the testimony of first officer Cassin's widow, they simply maintained that what Crosbie had professed *must* be true since "a knowingly untrue statement made by a witness to a royal commission on oath amounts to the crime of perjury." While Justice Mahon presumably regarded the Law Lords' easy assumption as naïve, if not disingenuous, he would have applauded their conviction that lying "on oath" before a Royal Commission constituted the crime of perjury.

In Macfarlane's view, "The most disturbing element of the Erebus affair is the Privy Council's economy with the truth." He cites in his book an authoritatively long list of erroneous claims the Law Lords made with respect to Justice Mahon, CEO Davis, Captain Gemmell, Captain Crosbie, and the Court of Appeal, among others, and suggests these economies with the truth were made not unwittingly but wittingly. The Privy Council's aim, he reckoned, was not merely to uphold the Court of Appeal's ruling on the extent of a royal commissioner's power and the requirements of natural justice; more sinisterly, it was also to find as *facts* "that no orchestrated litany of lies had ever existed, and that Air New Zealand's management carried no blame for the crash."

Despite multiple elisions in their report, the Law Lords' was a bravura performance in which they managed to award every party involved in the Erebus affair the most important part of what he wanted. Justice Mahon was declared correct in his "brilliant" analysis of the accident's causal factors. The Royal Commission's report "clears Captain Collins and First Officer Cassin of any suggestion that negligence on their part" contributed to the disaster. The aircraft had been locked on its nav track at the end of its outbound journey precisely because Captain Collins "was relying on the line he had himself plotted of the flight track" based on coordinates he'd obtained at the November 9 briefing. "It was a combination of his own meticulous conscientiousness in taking the trouble to plot for himself on a topographical chart the flight track that had been referred to at his briefing, and the fact that he had no previous experience of 'whiteout' and had been given no warning at any time that such a deceptive phenomenon even existed, that caused the disaster."

Chief Inspector Chippindale's attribution of pilot error was declared wrong—but, as a face-saver, only because the Privy Council in another economy of the truth claimed that he was unaware of certain important facts that were not discoverable at the time of his investigation but were by the time of Justice Mahon's. In fact, Chippindale had been acutely aware of the specified facts; he'd been the first to discover them. However, the chief inspector had elected to ignore this crucial data since none of it supported a finding of pilot error.

As for the Mahon Report's incendiary finding, in paragraph 377, that the airline's management witnesses had lied in concert in their testimony before the Royal Commission, the Privy Council rejected it out of hand. The Law Lords had no wish for anyone to imagine that Air New Zealand's welter of alarming administrative errors might have given rise to a coordinated attempt by the airline to conceal those errors. Instead, they contented themselves with noting for the record that some false testimony was only to be expected by witnesses swept up in a colossal accident investigation.

Such *individual* prevarications could charitably be attributed to human frailty and therefore readily overlooked.

In a memorable summation of the Privy Council's viewpoint, Mahon remarked that it was as if a group of singers had been discovered singing the same song but doing so "as soloists rather than as a choir." Using this rationale, the Law Lords upheld the Court of Appeal's finding that the royal commissioner had both exceeded his authority and breached natural justice's requirements. They themselves perceived no "orchestrated litany of lies."

The Law Lords frankly conceded that the airline's executive staff had committed innumerable unpardonable errors owing to the carrier's "slipshod system of administration and absence of liaison both between sections and between individual members" of its Flight Operations Division. These "appalling blunders and deficiencies"— *not* the pilots—they ruled responsible for the massive loss of life in a distant, frozen clime. After reaching this momentous conclusion, however, they seemed strangely unconcerned to address the grave defects the Mahon Report had revealed—and they themselves had candidly acknowledged—in the airline's operations. At the Royal Commission hearings, Air New Zealand's CEO had earlier conceded that "a series of events came together which resulted in a catastrophic magnitude," although he himself could not accept that "the company planning activity in principle was in any way inadequate." It now appeared that the Privy Council, like Morrie Davis, meant to insinuate that ineradicable human error was at the root of the tragedy.

Justice Mahon did not believe ineradicable human error had precipitated the fiery destruction of TE901 and all 257 individuals on board it. In his view, the accident resulted specifically from "the mistake made by those airline officials who programmed the aircraft to fly directly at Mt. Erebus and omitted to tell the aircrew." That error arose not because it was impossible to forestall all human errors, all the time, but because the airline's organizational procedures provided an opening for it to occur. Such openings could and should be closed by revamping Air New Zealand's wholly inadequate management structure. Only by learning the lessons of an aviation catastrophe and modifying industry practices accordingly could similar accidents be avoided in the future. In truth, Justice Mahon's highly productive investigatory approach, whose ultimate goal was maximizing passenger safety, resembles nothing so much as that of the inestimable NTSB.

The New Zealand government itself, in the form of counsel to its attorney general, made the case for dismissing Mahon's appeal during the Judicial Committee's hearings. At that time, counsel informed the august tribunal that "the New Zealand Court of Appeal is inevitably better placed than your Lordships" to judge the case. The attorney general "submits that the Court of Appeal's assessment should be given very great weight." In fact, "it would only be appropriate for your Lordships to differ from [the Court of Appeal decision], if you are quite satisfied that it is manifestly wrong."

Taking these sentiments to heart, the Privy Council found that it was indeed in the national interest—that is, in the interest of the government, judiciary, state airline, and that airline's regulator—for the Court of Appeal decision to stand. Deciding the merits of Mahon's appeal in favor of the appellate court's view of the extent of Mahon's authority and natural justice's requirements would obviously satisfy its five members. It would also conveniently relieve the Law Lords of any obligation to investigate the royal commissioner's conspiracy finding because it was beyond his powers to make it and, having made it anyway, he'd given senior airline witnesses no opportunity to hear and try to rebut it during the hearings themselves. "Very reluctantly" they then dismissed the case, benevolently declaring that "the time has now come for all parties to let bygones be bygones so far as the aftermath of the Mt. Erebus disaster is concerned."

The verdict was ironic. For one thing, unbeknownst to the Law Lords, Mahon had written a breakthrough report of international significance with respect to aviation safety, one that would forever change how transportation accidents are investigated around the globe. For another, this stellar judge of unquestioned brilliance and probity had a good-faith basis for concluding that a cover-up scheme had been hatched at the highest levels and was zealously being prosecuted by airline executives testifying before him at the Royal Commission hearings. That cover-up's purpose was to screen from public view the airline's serious systemic dysfunctions that were responsible for TE901's crash by aggressively attributing blame to the dead pilots instead. As required by the terms of his remit, Justice Mahon had then forthrightly shared his analysis. Finally, to those who knew him, Mahon *personified* natural justice.

PART FOUR

THE EMPTYING
HOURGLASS

17

KNIGHTHOOD?

AFTER THE JUDGE LOST THE SECOND OF TWO
high-profile legal actions—one involving the Court of Appeal and
the other the Privy Council—friends and admirers began to worry
about his welfare. Here was a man of tremendous intellect, energy,
and integrity who, as a distinguished member of the establishment,
had dared to challenge the position of New Zealand's prime minister, "his" national airline, and certain senior jurists during his official Erebus investigation. The Mahon Report was unquestionably
a masterpiece—substantively sound and elegantly written. That
its author had been deemed to have overstepped how he expressed
himself, which these supporters found unwarranted, was bad. The
creeping social ostracism that followed was not good either.

The independent-minded Justice Mahon, who did not recoil
from pitting wits with authority when confronted with high-
handed acts, had fully anticipated the negative reaction the New
Zealand government would have to his official accident report.
In an interview given five years after the loss of TE901 and all

its occupants, he good-naturedly explained how that body customarily operated in cases of scandal involving one or more of its agencies. If the findings of an official investigation were favorable, Mahon said, the government would officially embrace them. If the findings implicated one or more of its agencies, however, the government would summarily reject them. In the case of Erebus, the judge declared in the interview, the inquiry was "overshadowed by politics from start to finish, and I include the Privy Council hearing in that wide ambit."

Despite rebukes by ten judges in two countries, Peter Mahon was becoming increasingly revered by people from all walks of life throughout New Zealand. Justice Speight of the High Court (Sir Speight as of 1983) believed his good friend had "raised the image of the judiciary in the public eye by his courage and his outspoken attitude at a time when lesser (or perhaps more discreet) men would have remained subdued." Years later, a salty New Zealand radio personality would agree with this trenchant assessment. "The public sensed something special in him, some love of justice," declared broadcaster-turned-radio-host Paul Holmes, himself the author of a book on Erebus. Here was "a very brainy man who had worked out something very difficult and shocking, who had worked out why the DC-10 had crashed and had stood up to liars and bullies."

It was not uncommon for retiring or recently retired High Court judges to be knighted. This honor was wholly dependent on the favor of the government, of course, and no one expected Prime Minister Muldoon to look approvingly on a judge who had deeply and very publicly offended him. However, after the opposition party came into power in mid-1984 and installed a new prime minister, forty-eight eminent New Zealanders decided to present the case for Peter Mahon to receive a knighthood. The effort was coordinated by district court judge Anand Satyanand, who would later become the Right Honorable Sir Anand Satyanand and New Zealand's governor-general.

In his own letter of support, Satyanand identified three singular attributes of Justice Mahon. The first was "a laconic and mercurial brilliance, in both writing and speech, not often observed in this country." The second was "a judicial presence which was encouraging to the lawyer and litigant." The third, as recent events had abundantly attested, was a character "of bravery and principle."

Ted Thomas, then the president of the Auckland Law Society, wrote enthusiastically in support of the former royal commissioner, characterizing his service in the Erebus inquiry as "outstanding." "His conclusions have about them an inexorable logic," he noted admiringly of the man famed for his logician's mind. "Without his objectivity, determination and perception it must be doubtful whether the true causes of the accident would have been revealed."

That this was undoubtedly the case both Radio New Zealand's Carmel Friedlander and the *Auckland Star's* John Macdonald testified in their respective letters in support of a knighthood for Mahon. Having covered the Royal Commission hearings from start to finish, the two journalists were acutely aware of what the judge had been up against in his efforts to extract the truth from Air New Zealand's stonewalling officials. "His honesty would not allow him to stray from the truth for the sake of a popular verdict," Friedlander poignantly observed, which made him "as much a victim of the Erebus disaster as those who died in the crash." She had "boundless" admiration for "his courage and his honesty," as well as for "his fortitude and perseverance on behalf of his fellow men." What arrested Macdonald's attention during the hearings were the royal commissioner's "exhaustive efforts to seek out the truth from witnesses who were often secretive and hostile," his "ability to test experts" to "the very foundations of their knowledge," and his "double checking of evidence already submitted by the [government's] accident investigator" to obtain "new and vital clues to the inquiry." The man was indefatigable.

Richard Sutton, dean of the Otago University Law Faculty, similarly described Mahon's scholarly judgments as "models of lucid writing and profound research" that revealed an intimate grasp of the law's development over the preceding century. Mahon's handling of the Erebus inquiry's technical aspects Sutton singled out for unstinting praise. It was a "remarkable piece of analysis, showing to the full his ability to come to grips with expert evidence."

Recurring themes in the letters of support are Mahon's uncommon brilliance, attention to the finest detail, objectivity, integrity, bravery, tenacity, and fairness to all parties as well as a lively sense of humor and enviably lucid writing style. It must have come as a shock to the forty-eight letter writers when their petitions for a knighthood failed. The reason apparently was the intervention of Geoffrey Palmer, then serving as deputy prime minister, attorney general, and minister of justice in the new government. A mentee of Justice Owen Woodhouse, Palmer had reviewed for several hours the evidence before the Court of Appeal and concluded that, in their minority judgment, Justices Woodhouse and McMullin had it right, while the royal commissioner had it wrong. He could see no cover-up by Air New Zealand and was in any case totally unprepared to anger his mentor by endorsing a knighthood for someone that gentleman regarded with the utmost antipathy. "Justice Mahon was a very eminent New Zealander and he did a lot of good things," Palmer sniffed, "but [his report on Erebus] wasn't one of them."

18

LAST YEARS

THERE WAS ALWAYS SOMETHING PALPABLY NOBLE
and something markedly reserved about the former royal commissioner. Now there was something tragic as well. His health had
begun to erode.

Back in 1975, Justice Mahon had suffered a mild heart attack,
whose full implications perhaps were not then fully appreciated.
In the early 1980s, he was diagnosed first with heart failure (cardiomyopathy), then with a painful sinus tumor. The cancer progressed into his gum as well, but because of his heart condition,
no surgery would be feasible. Shrunken and gaunt, the former
royal commissioner was becoming a shadow of his former tall,
distinguished-looking self. Captain Jim Collins's widow, Maria,
who possessed vast reservoirs of respect for Mahon, visited him
on several occasions. "He was always quite difficult to get verbally
close to," she would later recall. Even toward the end there thus
remained "a certain distance." When Justice Speight visited his old
friend during his last illness and inquired how things were, Mahon

poetically responded, "I nightly pitch my moving tent a day's march nearer home."

The path of up-and-coming writer Greg McGee crossed with that of Justice Mahon in 1985, a year after the judge had published *Verdict on Erebus*. It happened as the result of an offer McGee received one day from the head of drama at Television New Zealand (TVNZ). Would he adapt Mahon's critically acclaimed book into a powerful four-hour miniseries?

McGee surmised that he had received the offer in part because he had a legal background; obviously, anyone tackling the acrimonious topic of Erebus for public television would benefit from having this sort of expertise in addition to a talent for scriptwriting. McGee's initial talks with producer Caterina De Nave did not go well. She appeared oddly reticent to throw herself into an explosive story whose whole purpose was to dramatize recent corruption in the corridors of power. McGee wondered if she already was experiencing outside pressure. He also objected to the limited scope of the envisioned project. *Verdict on Erebus* ended with the conclusion of the Royal Commission of Inquiry, which, in his view, could not provide enough interest to sustain a four-hour series. TVNZ's head of drama encouraged McGee to go meet Peter Mahon himself and talk it over.

Upon arriving at Peter and Margarita's home, McGee was "stunned by how emaciated Mahon was." Only a few years earlier, the judge had been royal commissioner and popped up constantly in the television news. At that time, he was tall and athletic looking, with a patrician bearing and "penetrating eyes and saturnine jowls of an elegant bloodhound." But now he'd been much reduced by a series of medical problems that his wife attributed to stress over what had happened in the wake of the Royal Commission of Inquiry.

From the ill judge, his voice reduced to a whisper, McGee got an overview of what had transpired since he had released his official accident report on the Erebus disaster. For starters, there was the imperial Prime Minister Muldoon's unprecedented rejection of the Mahon Report's findings. This was soon followed by what amounted to a defamation suit by Air New Zealand against Justice Mahon. From the judge, still speaking sotto voce, McGee gained insight into his brave fight with the country's government and judiciary. McGee returned to TVNZ totally energized by the enlarged story he wanted to tell. That bigger story would be about "how a Judge was commissioned by the government to find the truth—and what happened to him when the truth he found was unpalatable to that government." In a nutshell, it would be about "Mahon's quest and ultimate tragedy."

"It's easy to forget the atmosphere of fear Muldoon engendered," writer Greg McGee reminded readers in a chapter of his memoir recounting how he came to agree to script *Erebus, The Aftermath*. The envisioned docudrama had an outstanding independent legal advisor in Gary Harrison, junior counsel assisting Justice Mahon at the Royal Commission of Inquiry, and McGee himself possessed a legal background. Nonetheless, he was concerned that "most of the characters in this drama were still alive and that some of them were highly litigious." McGee thus insisted on being indemnified against any defamation action by having TVNZ—not his own writing team—assume responsibility for ensuring that none of the material used in the miniseries could be defamatory. McGee also saw fit in his negotiations with TVNZ to request security for him and his family when the completed miniseries eventually aired over a period of four nights. It was *that* volatile and litigious a time.

Acquiring as much new intelligence as possible on the Erebus accident's stormy aftermath was an essential component of McGee's vision for the docudrama-to-be. From the first, however, producer De Nave balked at the idea of a bold investigatory approach. Justice Mahon had earlier cut out about a third of his draft for *Verdict on*

Erebus at the urging of lawyers that considered them potentially defamatory. McGee was alarmed that his new collaborator had zero interest in their reading what had been excised to enable the two of them to form their own view of the trustworthiness of the material.

Even more inexplicable was De Nave's taking the trouble to fly to London "for research." There, visiting the Privy Council's quarters, she uncovered the most jaw-dropping intelligence possible for an Erebus researcher—and never shared it with McGee. That did not mean she didn't share the shocking revelation with anyone else, however. The elite few in the know included first officer Cassin's widow, Anne, who in 2015 reported that "when the producer Caterina De Nave interviewed me in St. Heliers . . . for *Erebus, the Aftermath,* she told me that she had gone to England to try to find evidence. She found out that two of the Appeals Court Judges with connections to Air New Zealand had been flown by Air New Zealand to England," where they met, separately, with the Law Lords.

Despite her astonishing failure to share a breathtaking discovery with McGee, he shrewdly figured out for himself what must have been the situation. His clue was the Privy Council's decision. It was "so contorted that I'm left wondering if someone had a quiet word with their lordships that it was in the national interest for the CA [Court of Appeal] decision to stand."

Anyone attempting to understand the air disaster on Mount Erebus and its protracted and highly rancorous consequences suffers initially from evidence overload. Being no exception, McGee plodded through copious documents "not knowing enough to venture an opinion." While others on the docudrama team were out interviewing virtually everyone at Air New Zealand, McGee personally conducted all talks with Justice Mahon. He worried about what would happen if in the end he believed the judge had been wrong. But, slowly, random pieces of a complicated jigsaw puzzle started to fall into place. Finally, McGee was able to tell Mahon that "it was clear to me that a) the witnesses were lying, b) they were lying in concert, and c) to achieve that, they must have put their heads together before taking the stand."

In the end, McGee concluded that Justice Mahon's analysis of the cause of and culpability for the accident was "between 80 and 100 percent right." Under normal circumstances, that sort of achievement—especially in a case as "difficult, complex, emotionally charged and technically obtuse" as Erebus—would have been widely admired by his fellow judges. Instead, a number of colleagues broke with Justice Mahon, the Court of Appeal's Justices Woodhouse and McMullin being conspicuous cases in point. In McGee's view, their minority judgment was downright vicious, "about as scathing of Justice Mahon as legal decorum allows." The two caused the judge deep personal distress with their imputation that, despite months of sizing up witnesses during the Royal Commission hearings, he had no grounds on which to reach the findings he did regarding their credibility.

During one of their meetings, Justice Mahon told scriptwriter McGee that he believed the government had made him royal commissioner "because it thought Brown could exploit their close relationship," thereby ensuring that the judge would be a safe pair of hands. During another session Justice Mahon confided, in "his raspy, weakening voice," that he knew the identity of the orchestrator of the notorious litany of lies. Justice Mahon had given considerable thought to the matter because he did not consider Air New Zealand CEO Davis capable of personally organizing the grand deception. Greg McGee was of the same opinion. The way he put it was that Davis "didn't actually know enough about the operations side of the airline to run such a sophisticated technical defense." He was merely a "reliable mouthpiece for those who stood behind him."

But who stood behind him? It was Prime Minister Muldoon's private lawyer, Des Dalgety, Justice Mahon murmured. Dalgety, whose office was on the nineteenth floor of New Zealand House, was the orchestrator. This made perfect sense to McGee in that Dalgety was both the prime minister's personal attorney and director of Air New Zealand's board. As such, he made the perfect liaison to communicate Muldoon's directives on strategy to Air New Zealand and to bring back from someone in the airline's operational command "the tactics necessary to execute" those directives. Although McGee mentioned no name, the "someone" in the airline's operational command would surely have been Captain Gemmell, the airline's chief pilot–turned–flight manager (technical), who was the behind-the-scenes architect of the Antarctic flights and, after the crash, day-to-day manager of Air New Zealand's case.

During Justice Mahon's final year of life, 1986, McGee's literary project progressed steadily. Only when the final drafts needed to be signed off on did TVNZ's director general finally grasp the potential legal risk the docudrama would pose. Reportedly "shit scared," he demanded that one of the country's top defamation attorneys vet the scripts—resulting in half the docudrama's contents being ruled out as potentially defamatory. Fortunately for McGee, his team had kept meticulous source scripts. These "recorded the origin of every bit of information, every scene and virtually every line of dialogue attributed to any character." The information proved a godsend; every questionable item except one was quickly restored to the script. Actors were told that if they stuck to the scripted dialogue, they'd be protected by TVNZ if any defamatory action was later launched.

By the time *Erebus, The Aftermath* began airing its four segments on October 18, 1987, Justice Mahon would be dead. McGee was relieved because it spared him "the burden of once more standing up and defending himself and his integrity against the likes of Morrie Davis."

Breaking all New Zealand viewing records, the docudrama proved a stunning success, in part because it boasted a cast of outstanding actors. In addition, as one reviewer put it, *Erebus, The Aftermath* convincingly portrays "the devious nature of the Muldoon Government's handling of the inquiry that destroyed the reputation of Peter Mahon, the judge appointed to head it." Predictably, Air New Zealand's executives excoriated the docudrama and reportedly threatened TVNZ with a lawsuit if it ever aired the miniseries a second time. Somehow, it would one day wind up on YouTube.

19

FUNERAL

PETER THOMAS MAHON, AGED SIXTY-TWO, SLIPPED away in the early morning hours on August 11, 1986. His still body was found by his wife, who had gotten up to make a cup of tea. As she would later recall, "I just sort of peeped into his room," then "quietly crept forward," aware a trifle could awaken him. Mahon hadn't eaten the tiny, soft sandwiches she routinely left in the refrigerator for him to snack on during the night. "Peter? Peter, are you all right?" she whispered.

Margarita Mahon had desired a private funeral for her husband, but it was not to be. Throngs descended on the Holy Trinity Anglican Cathedral in Auckland's historic, upscale suburb of Parnell, all eager to pay tribute to the brilliant jurist unafraid to break ranks and speak truth to power. Justice Mahon had regarded as his sacred duty to uncover the cause or causes of an unimaginable accident that cost 257 airborne sightseers their very lives. His responsibility, he felt, was to them, the deceased—not, as the prime minister and the CEO of Air New Zealand would have it, the defense of the airline at any cost. So thorough-going was the judge's demolition of the airline's case against the pilots and so persuasive was his own case *for* the pilots that the equivocal Privy Council could only applaud his painstaking, multifaceted, substantive investigation—even if that body *was* prepared to be lax with Air New Zealand's executives owing to human beings' tendency to lie (singly, not jointly) when cornered.

Even Attorney General Jim McLay, reputed (despite appearances) to have been a stout defender of the judge, was seen to shed tears at the funeral. Did he, Sam Mahon wondered, regret having "run with the pack" to badger his father for exposing inconvenient truths? Although on cordial terms with Muldoon, the polished and urbane McLay has conceded that the two never became close. Justice Mahon's son personally concluded that the attorney general was fundamentally "a good man coerced."

Both at the funeral and subsequently in interviews and a book, Sam Mahon shed a clarifying light on the steadfast character of his emotionally reserved father. Speaking at Auckland's Holy Trinity Anglican Cathedral, he suggested that Peter Mahon "defended legal principle in the same way that Sir Thomas More defended it, as a lattice on which we hang civilized society." Like the earlier "man for all seasons," this later one revered the law and trusted in the law's ability to defend against injustice. "The somber experience of mankind, correctly identified by the church and by the law," the senior Mahon once observed, demonstrates that "the generality of men abstains from breaking heads, and from breaking contracts, through fear of legal retribution, and that retribution is exacted by the law as an organ of society and with the approval of society."

Like More, Mahon possessed—besides intelligence and integrity—the resolute determination of a person who cannot be tempted to forsake himself, even to save himself. Yes, the judge was prepared to yield inessential parts of himself to the encroachments of others. However, when it came to surrendering his essential core, he reacted just as More had—he "set like metal, was overtaken by an absolutely primitive rigor, and could no more be budged than a cliff." For Mahon, putting out a false message about the real circumstances of TE901's horrific collision with Mount Erebus would never have been an option: how could he possibly have compromised his own view of himself as a staunch defender of the law? Even when, as had happened to More, Mahon witnessed the law being subverted before his very eyes, he didn't blanch. Like More, he persisted in taking advantage of all available legal remedies open to him. Those included innumerable public interviews and addresses, a confident appeal to the Privy Council, and the triumphant publication of *Verdict on Erebus*. The book sold out in a week, went immediately into reprint, and won the New Zealand Book Award for Non-Fiction in 1985.

It was not a foregone conclusion that artist-activist Sam Mahon would prove a gifted interpreter of his distinguished father's character. The two men were cut from very different pieces of cloth, which made the young rebel-in-the-making's understanding of his distant but tolerant parent tenuous at best. Nonetheless, the judge had intensively exposed his son to certain moral principles during his formative years. As a youth, Sam had sensed that it would be highly imprudent for him to challenge any of them by defiant behavior on his part. These credentials would one day make Justice Mahon's elder son a particularly insightful commentator on the clash between the prevaricating airline's executives and the pugnacious prime minister on the one hand and the truth-telling royal commissioner on the other. What's more, Sam understood his father's "obstinate refusal to back down in the face of adversity" because he himself admitted having acquired this trait from extended exposure to him.

According to Sam's remarks in a variety of venues, Justice Mahon prized as the core aspect of his judicial self his ability to distinguish truth-tellers from liars, and he had no use whatsoever for the latter. Sam himself had "learned as a child never, ever to" attempt to deceive his father. "He was a tough man. There was very little tolerance." Had Sam, growing up, told whoppers like the airline witnesses had relentlessly done before the royal commissioner for weeks on end, he knew he would have been cut to shreds. "My father drew very, very severe lines." Any sustained, coordinated effort to give the runaround to this righteous judge was thus bound to end in fireworks.

In Sam Mahon's view, the law was a good fit as a calling for his father. "It suited his wit, [and] it suited his sense of gamesmanship as well, for he was a strategist and a card player." The judge's natural taciturnity would prove a considerable asset for him during his Erebus investigations, which he apparently regarded as a game of chess. With great care, "he quietly played the pieces" in an increasingly fraught contest. His final move, made in the privacy of his own home, was to coin the expression "an orchestrated litany of lies." That memorable phrase was intended as a rebuke of the Air New Zealand and Civil Aviation executives, together with their legal representatives, for fabricating a false tale about TE901's crash in their testimony before him at the protracted Royal Commission hearings.

Justice Mahon knew from the outset that the Erebus case would be a political football, but he revered the law, wished to do right by those who perished on the mountain, relished a challenge, enjoyed being in the spotlight, and could not be intimidated. The price the judge would pay for his brilliant and courageous report proved to be steep: his career, cherished friendships, financial security, and possibly the foreshortening of his life. While it is true Peter Mahon had developed a weak heart prior to his appointment as royal commissioner for the Erebus inquiry, it is also true that he thrived on cases involving conflict and adversarial competition. Maybe sixty-two years was all he had, son Sam suggested philosophically. As for Erebus, far from being the end of his father, Sam believed it was actually "the *making* of him." The judge was "an artist in his way," the artist-son thought, "and Erebus was his masterwork." To Sam, "it is inconceivable he would have turned down the challenge of untangling that extraordinary mystery."

Justice Mahon was armed only with his integrity, his belief in the law, and his hard-won discovery that Air New Zealand and Civil Aviation—not the pilots—were responsible for the deaths of a planeload of adventuresome sightseers. He had nonetheless taken on powerful individuals in high places irate he would first uncover and then expose two inconvenient truths. The first was that serious systemic issues in the national carrier's dysfunctional Flight Operations Division, along with negligent oversight by Civil Aviation, had caused TE901's crash. The second was that a no-holds-barred conspiratorial effort had been made to hide the airline's culpability for the accident by blaming the dead pilots instead.

Paul Davison, a future Court of Appeal judge who as a young man served as the Collins family's counsel and NZALPA's assisting counsel at the Royal Commission hearings, would in later life describe the strong influence the extraordinary judge exerted on him. Despite the enormous pressures under which the royal commissioner labored, Davison watched him courageously and adroitly do what was right, using the law as a tool "by which justice can be achieved if you adhere to the process." Step by methodical, logical step Mahon slowly stripped the government's case of pilot error of any credibility whatsoever and established the real causes of and responsibility for TE901's horrific collision with an Antarctic volcano. To Davison, who keeps a framed photograph of Mahon on his bookcase, he was "an example of the law working."

PART FIVE

THE LONGER
VIEW

20

MORE LIVES LOST, 1980–1989

SOMETIMES OUR ACTIONS HAVE UNINTENDED CON-sequences. In the view of the International Civil Aviation Organization (ICAO), the New Zealand government's indefinite shelving of the Mahon Report proved to be one of these. As a result, the important "message from Antarctica," which the organization deemed well ahead of its time, was not immediately heard. In ICAO's view, that was most unfortunate because shelving the Mahon Report caused a flurry of potentially avoidable transportation-related disasters to occur during the 1980s. The last of these—the accident at Canada's Dryden Municipal Airport in 1989—was eerily reminiscent in its dynamics to the Erebus crash.

That certain thorny issues invariably come with the job of scrutinizing a carrier and its regulator for possible gross negligence is only to be expected. The Dryden accident illustrated this truism as vividly as the Mount Erebus catastrophe had earlier. Presiding over the public inquest into the Canadian disaster, Justice Virgil Moshansky thus immediately found himself, like his New Zealand predecessor, up against a hostile group of aviation industry witnesses.

Superficially, the Dryden accident had the hallmarks of pilot error. Declining to interpret his mandate narrowly, though, Justice Moshansky took the same expansive view that Justice Mahon had, deeming it essential in the interests of aviation safety critically to review salient aspects of the organizational arrangements of Air Ontario, a subsidiary of Air Canada. Finding them as sorely wanting as the Kiwi judge had found Air New Zealand's, he brilliantly traced the roots of the disaster back to systemic errors, the study of which Justice Mahon and Captain Vette had pioneered during the Erebus inquiry and to which they had attributed the loss of TE901 and all its occupants. Because Justice Moshansky did not know that the Mahon Report existed, however, he had to work out for himself how organizational accidents occur before submitting his 1,700-page report implicating the entire Canadian air transportation system.

During the public hearings, Air Ontario took as belligerent a stance as Air New Zealand had toward what was intended to be a neutral, fact-based investigation into the causes of an unacceptable loss of life. In the Canadian case, the disaster happened less than one minute after the accident aircraft took off in a snowstorm. Through counsel, the airline attempted to limit the scope of the inquiry and brought an action to stop Justice Moshansky from naming in his report anyone whose shortcomings had been found to have contributed to the tragedy. While the action didn't prevail, resisting it cost Justice Moshansky much time and wasted taxpayers' money in the same way Air New Zealand's obstructionism impeded the timely conclusion of Justice Mahon's case (hence the latter's costs order against Air New Zealand). Although Justice Moshansky *was* allowed to name those at Air Ontario who were implicated in the Dryden accident, the president and CEO of the carrier never conceded that it was totally preventable, notwithstanding enormous amounts of expert testimony to the contrary. This, of course, was also true of Air New Zealand CEO Morrie Davis.

Justice Moshansky fared somewhat better than Justice Mahon when it came to dealing with the testy regulator of the airline under investigation. Similar to what Mahon had uncovered about the Ministry of Transport's Civil Aviation Division, Moshansky found the regulator in his case guilty of "serious policy lapses, inconsistent policies, bureaucratic bungling and human failures." To that regulator's "great credit," however, Moshansky noted three years after the release of his report with its 191 recommendations, it had quickly started to implement the proposed changes.

That had not been Justice Mahon's experience with Civil Aviation, which was as determined as Air New Zealand itself to deny any responsibility for the Erebus tragedy. Before the Mahon Report's release, not too much attention had been focused on whether the airline's regulator might have some potential liability for the disaster. Overnight, the Mahon Report changed the prospects for Civil Aviation. The implications were especially serious because, as part of the government, Civil Aviation would face unlimited liability if found culpable.

Already at the Royal Commission hearings themselves, one of Civil Aviation's longtime employees (recently retired) had angrily sensed which way the wind was blowing and set out to write a slim book devoted to the defense of Civil Aviation's conduct with respect to Air New Zealand's Antarctic flights. Titled, *The Erebus Enquiry: A Tragic Miscarriage of Justice*, C. H. N. L'Estrange's tract was essentially a paean to Chippindale's findings, which exonerated both the airline and its regulator of any responsibility for the catastrophe. It was also a denunciation of certain ideas emerging at the Royal Commission hearings that eventually found their way into the Mahon Report. Civil Aviation itself never backed down from its unyielding stance. Years later those in its orbit would still be reviling the Mahon Report.

Justice Moshansky did possess two decided advantages over Justice Mahon in his accident investigation. One was that he did not need to contend with the existence of an antecedent in-house government report that had effectively already cleared everyone at the airline and its regulator of any responsibility for the disaster in question before his own inquiry could even commence. The other was that he did not have Robert Muldoon as his prime minister.

The minds of both Prime Minister Muldoon and Air New Zealand's CEO, Davis, were completely closed to the possibility that systemic or organizational error played a causative role in the accident on Mount Erebus, which explains why the Mahon Report got shelved. Successive administrations saw to it that the report stayed shelved for a full two decades. Perhaps it is poetic justice that the man responsible for the Mahon Report finally being tabled in Parliament in 1999, making it an official document at last, was the government's own transport minister, Maurice Williamson.

Working at Air New Zealand for an extended period (1975–1987) that included the collision of TE901 with Mount Erebus and its fraught aftermath, Williamson had been, so he would put it years later, "deeply involved in what went on." Presumably this was hyperbole—an allusion to his witnessing, not participating in, the wholesale destruction of evidence ordered by the airline's CEO, Morrie Davis, in its chaotic wake. When the Mahon Report was released, longtime Air New Zealand employee Williamson started "dancing around the Maypole," so excited was he that New Zealand had "somebody the caliber of Mahon." But then nothing happened—for years—as multiple administrations declined to table in Parliament the royal commissioner's breakthrough analysis of what caused the crash of TE901, where responsibility for it lay, and what steps should be taken to prevent any future accidents of this type.

Convinced that "Justice Mahon got it right," Maurice Williamson, who was later elected to the New Zealand Parliament, was able to act once he was appointed transport minister two decades after the air disaster. "It was so wrong to blame just the pilot," he would explain, when "all of the systems" had failed him. "It was a systemic failure, not one error." Looking back later, at the end of a long political career, Williamson would consider the tabling of the Mahon Report one of its greatest highlights. At the ceremony at which he did so, the minister of transport lauded Justice Mahon as "one of New Zealand's true heroes" and declared that his seminal report richly merited governmental recognition. Present in Parliament to witness the event were the widows of Captain Collins, first officer Cassin, and Justice Mahon.

Conspicuous by her opposition to Minster Williamson's action was a furious member of Parliament named Robyn McDonald. She was the daughter of Captain Kippenberger, who had been head of Civil Aviation, the governmental body with supervisory responsibility for the Antarctic flights' safety, at the time the unimaginable had happened. Not surprisingly, McDonald did not wish to see a report that revealed Civil Aviation's negligence tabled in Parliament. To mitigate the reputational damage that she supposed Williamson's act would do, Kippenberger's daughter attempted to get the Court of Appeal's two judgments along with the Privy Council's findings tabled alongside the Mahon Report. She was denied.

What Maurice Williamson and others did not appreciate the day of the ceremony in 1999 was that giving the Mahon Report official recognition in New Zealand did not automatically ensure that the government would forward it to the International Civil Aviation Organization (ICAO), of which the country was a member. Only in 2012 would ICAO formally confirm that Justice Mahon's breakthrough analysis of TE901's crash was in its possession. However, the Mahon Report is not to be found within Annex 13, the go-to resource for the global aviation community when it wants to learn about a particular aircraft disaster and investigation report associated with it. Annex 13 holds only official reports written in compliance with ICAO's specific protocols for it. For this reason, the Mahon Report is not included there, while the earlier Chippindale Report is. That report's author remained in high standing at the international body. He continued for years to be actively involved in aviation assignments on ICAO's behalf.

Once ICAO *was* belatedly in possession of the Mahon Report, its members were instantly impressed by the theoretical breakthrough that Justice Mahon had achieved working collaboratively with Captain Vette. In the Erebus crash, as the report made clear, latent and active failures combined to precipitate an "organizational" transportation accident. The prestigious international body deemed Mahon and Vette's analysis "probably ten years ahead of its time." So important was it that ICAO went so far as to declare that "in retrospect, if the aviation community and the safety community at large had grasped the message from Antarctica and applied its prevention lessons," the world's transportation disasters of the 1980s would not have occurred. ICAO then specifically identified the Bhopal gas tragedy at Union Carbide Limited's pesticide plant in India (1984), the Chernobyl nuclear accident in Ukraine (1986), the King's Cross fire at a major interchange on the London Underground (1987), and the Clapham Junction train crash just south of the Clapham Junction railway station in London (1988) as among the horrors the world could have been spared.

As for the Air Ontario accident (1989), ICAO declared by way of conclusion, "certainly the Dryden Report would not have existed." There would have been no need for a Dryden Report because, thanks to the Mahon Report, the formula for avoiding such a catastrophe was already available. Unfortunately, the formula was inaccessible—on the shelf, so to say.

21

THE NATURE OF
COURAGE

THE CHIPPINDALE REPORT INITIALLY SO SWAYED popular opinion that it almost became the accepted verdict in New Zealand on what caused the Erebus disaster and who was responsible for it. That it did not was owing to the prodigious intellectual labors and enormous political courage of two towering obstacles. They were the High Court's incorruptible Justice Peter Mahon and the principled top Air New Zealand pilot Gordon Vette. The former's mission was to determine the accident's dominant cause, which the latter's investigation into the tragedy's proximate cause served to substantiate.

In *Profiles in Courage,* President John Kennedy wrote, "A man does what he must—in spite of personal consequences, in spite of obstacles and dangers and pressures—and that is the basis of all human morality." It definitely was the basis of Justice Mahon's morality—and Captain Vette's as well. In the judge's case, his uncommon psychological constitution, which enabled him unswervingly to follow the dictates of duty as he understood them, may have been partially inherited. When news broke that the Mahon Report's publication was creating an uproar, one of the judge's two sisters would later recall thinking at the time, "it seems like history is repeating itself." When the three Mahon siblings were growing up during the Great Depression, their father had managed a general store. One day the senior Mahon was shown a shady bookkeeping trick to fudge the numbers and urged to start using it. He refused and promptly lost his job. The father had suffered *then* for doing what he considered to be right. The son was prepared to suffer *now* for doing likewise. Both men abhorred anything not "straight" and true.

Sam Mahon spoke movingly about his father's incorruptibility when it came to protecting the rule of law, no matter the professional or personal cost. "My father was the bravest person I have ever met." According to Sam, the judge would "never be pushed, never give way, never compromise" despite the enormous pressures undoubtedly put on him, during the highly charged Erebus-related legal proceedings, to do so. "They picked the wrong man to try to squeeze."

Author James McNeish agreed with Sam's assessment in his posthumously published book on three prominent Kiwis, of whom Justice Mahon is one. The judge was, most conspicuously, capable of breaking ranks and taking on members of his own professional fraternity if he discovered something so odious that his conscience would not permit him to do otherwise. This singular trait ensured that Peter Thomas Mahon would himself become a victim of Erebus. While conceding that her husband did have choices, Margarita observed that he *was* badly treated by some of his peers. "I see it as a great blot to [*sic*] the New Zealand legal system."

There are still those today who consider the former royal commissioner a flawed hero. They are honorable men of high

standing—prominent members of the legal fraternity mainly—
and they were pained witnesses to the rancorous efforts of a minority of their brethren to give a supremely gifted and esteemed
colleague what euphemistically might be called "a comeuppance."
They wish that he had not called things as he saw them since, in
their apparent view, no one really needed to hear that the airline
had committed a criminally long list of errors *and then* tried to
cover them up by displacing responsibility for the world's fourth-
worst aviation accident onto the shoulders of the deceased pilots.

But blaming two outstanding airmen with exemplary flight
records for the carrier's own mistakes was precisely the sort of
inequity Mahon had spent a lifetime abhorring. Given the startling facts surrounding the accident that he'd uncovered, the judge
simply was not prepared to reach such a finding. His duty was to
investigate, then to report back with the results of his investigation.
The evidence so clearly indicated an organized plan of deception,
Justice Mahon wrote a correspondent in Europe, that "I really had
no alternative."

The Privy Council might have fancied Mahon a naïve colonial
judge unable to grasp his unspoken duty to the New Zealand establishment when conducting an accident inquiry. But those with
integrity and possessed of better judgment consider the incorruptible Mahon to be a heroic man for all seasons. Through months of
trying Royal Commission hearings, he held the scales of justice in
his steady hands. Then, in the fullness of time, Justice Mahon delivered the flabbergasting truth to his fellow citizens. It has just taken
a horrified country a good, long while to begin to understand that.

According to his wife, Justice Mahon believed that he had executed his task as royal commissioner as ably as he could and even
if given the chance, would not have changed his credibility finding.
What gravely disillusioned him was the biased, "scathing" criticism
of his legal abilities that the Court of Appeal's Justices Woodhouse

and McMullin had delivered in their minority judgment. It's reported that, as his own end approached, Peter Mahon often dwelled on these words from *The Divine Comedy's* "Inferno" in connection with his Erebus experience:

> Where the instrument of thinking mind
> Is joined to strength and malice
> Man's defense cannot avail
> To meet those powers combined.

Captain Gordon Vette is a second exemplar of what President Kennedy admired most in others—political courage. Even those at Air New Zealand who knew the flight track had been misprogrammed just hours before the fatal flight initially were at a loss to explain how Captain Collins and first officer Cassin could plow straight into a mountain right in front of them in broad daylight. This mystery may have given Chief Inspector Chippindale's team the inspiration to craft an official report that shifted blame for the accident away from the airline's organizational deficiencies and onto the dead pilots' alleged loss of situational awareness—that is, inattention to their dangerous surroundings. This theory was proved wrong by Captain Vette, who risked his highly distinguished career to go up against his own company's formidable CEO to discover the real reason why the pilots failed to see the mountain before smashing into it. It was because sector whiteout on the early afternoon of November 28, 1979, prevented the rising slope of Mount Erebus from being visible to human eyes on approach to Lewis Bay.

For establishing that the proximate cause of TE901's crash into Mount Erebus was not pilot error but sector whiteout, Captain

Vette was forced out of Air New Zealand with "seven years to run as their top pilot." Although Air New Zealand CEO Davis made his expulsion as painful as possible, the resourceful Captain Vette managed to reinvent himself and enjoyed a second career in which he passionately promoted aviation safety internationally in multiple ways. By 1998, when the University of Glasgow awarded him an honorary Doctorate in Engineering, Vette's accumulated good works both personal and professional were earning him praise as an "outstanding role model" as well as "a man of undoubted integrity, a highly trained and gifted pilot, with an inquisitive, deductive brain of high intellect."

Vette's contributions to aviation safety were practical as well as conceptual, as the University of Glasgow appreciated in honoring him in the field of engineering. He was, for example, an early and earnest advocate of the creation of the sort of forward-looking ground proximity warning system that could have saved TE901 had it been available in 1979 (as noted, Collins and Cassin had had only six seconds of warning in which to avoid the mountain). Continuing his research on crashes into terrain, Vette succeeded in accelerating the development of such a system. By 2007 terrain awareness and warning systems (TAWS) had been installed on 95 percent of commercial jets worldwide.

Twenty years after the crash of TE901, at the same convocation in Auckland at which Justice Mahon was honored posthumously for *his* contributions to aviation safety, the International Federation of Air Line Pilots' Associations awarded Gordon Vette a Presidential Citation in recognition of and appreciation for his invaluable work in the field of aviation safety. There followed a further public acknowledgement of his contributions in this area in the 2007 Queen's Birthday Honors, on which occasion he was appointed a member of the New Zealand Order of Merit. Several years earlier the seventy-three-year-old had suffered a serious stroke, which deprived him of the power of speech but mercifully left all his mental faculties intact. Mark Vette called his father's appointment "a final congratulations from the country."

A different variety of courage was exemplified by the elite cadre of New Zealand police officers whisked to the crash site the day after the accident to conduct a harrowing body recovery effort. Chipping disfigured human remains out of the ice to enable their ultimate return to the victims' families back at home may have been a noble endeavor, but it was fraught with exceptional physical and mental risk. The disaster victim identification (DVI) squad and their search and rescue associates might easily have perished in the dangerous work area from one of several perils although, thankfully, none did. They might just as easily have developed long-term psychological ramifications from their nightmarish ordeal of handling an unfathomable number of dead bodies—and Constable Leighton was one of those who did later receive a diagnosis of posttraumatic stress disorder.

When the police officers returned to New Zealand from their grim Antarctic mission, there was no immediate public acknowledgment of all they had risked during the body recovery operation. Only on November 28, 2006, the twenty-seventh anniversary of the Erebus disaster, did Prime Minister Helen Clark announce that a New Zealand Special Services Medal (Erebus) would be awarded to a range of organizations and agencies involved in Operation Overdue. The crash of TE901 was "without question the bleakest moment in the history of New Zealand's fifty-year presence in Antarctica," she declared. The harrowing work of those associated with the operation's body retrieval phase "far exceed[ed] the boundaries of what could be expected in the course of normal police, search and rescue, or air accident investigation duties." Yet the dangerous, traumatic effort was critical both to assuage the grief of the victims' families and to help the country come to terms with a disaster causing death on such a grand scale. Between both phases of Operation Overdue, 214 of 257 accident victims were eventually identified. At that time, the achievement set an international standard and model for victim identification procedures in mass tragedy events.

Further public praise of the body recovery effort followed when flight engineer Gordon Brooks's niece learned about the police officers' mission on the mountain from her cousin and decided to produce a documentary on it. That remarkable work, known as *Erebus: Operation Overdue* as well as *Erebus: Into the Unknown*, was released in 2014 to great acclaim. One reviewer astutely described it as "a hellish story told hellishly well." The film featured a physical reconstruction of the work area authentic in its details, reenactments, and moving interviews with the four police officers who had earlier received the Special Services Medal (Erebus). In the film, Sergeant Gilpin and Constable Leighton both excelled at recreating the fear and trepidation that accompanied them as they jumped out of a helicopter into blindingly white nothingness—and their ordeals subsequent to that first one.

In addition to physical courage, Sergeant Gilpin and Constable Leighton possessed a type of moral courage that was shared by first officer Cassin's feisty widow, Anne. The police officers were disturbed belatedly to discover that the contents of Captain Collins's ring binder had mysteriously disappeared sometime after they'd found this key piece of evidence at the crash site. After meticulously following the protocol DVI officers had been instructed to use when turning in finds, they found it incredible that the ring binder's perfectly legible contents were now being said to have been destroyed because they were "illegible." Gilpin, in particular, made a ruckus about the incident for years afterward. In Anne Cassin's case, what disappeared was not pages in a ring binder but pages in a folder her husband had been given at the pilots' briefing session and had left at home by mistake the morning of the accident flight. If made public, either set of pages would have proved *conclusively* that Captain Collins had been instructed that his flight path would be over McMurdo Sound's flat sea ice, not over a towering volcano on Ross Island. If such incontrovertible evidence had reached the public, the airline might then have felt constrained to attempt salvaging the situation by accepting responsibility for the disaster, making the indicated payouts to the families of victims, and undertaking to eliminate its dangerously dysfunctional operational setup by implementing Justice Mahon's sound recommendations.

None of that transpired. Instead, with the backing of an uncompromising prime minister, the equally resolute national airline chose a strategy of destruction of all incriminating evidence, wherever it might be. In the near term, this allowed the national carrier to appear innocent of gross negligence. In the longer term, it afforded airline executives the breathing room in which to conspire for the purpose of mounting a multiphased cover-up of the carrier's responsibility for a colossal disaster. Although relieving the pressure on Air New Zealand for a while, these maneuvers thrust the grieving families of Erebus victims into a vortex of confusion over what really happened to rob their loved ones of their lives.

⎯⎯⎯ ◈ ⎯⎯⎯

Among the grieving families, none faced more painful prospects than those of Captain Jim Collins and first officer Greg Cassin since the airline's plan was to blame the two pilots for the world's fourth-worst air disaster. First officer Cassin's widow, Anne, was unusual in having a solid background in aviation herself, which had commenced in childhood under the tutelage of her father—pilot and instructor John Munro. When he was killed in an accident shortly after takeoff in November 1965, she continued her flying instruction under one of Munro's junior instructors, Greg Cassin. Soon after the two married, Anne got her private pilot license (PPL). At some point, Greg joined Air New Zealand. It is hard not to imagine that, for Anne, November 28, 1979, would have been the worst day of her life. Some months later, to add insult to injury, she learned from Chief Inspector Chippindale's interim report that the *pilots* would be deemed to have caused the catastrophe.

Anne Cassin's suspicions had been quickly aroused by the individual deputized by NZALPA to liaise with both widows of the accident flight's pilots. Captain Crosbie's probing questions about her husband's briefing documents, which he had forgotten to take with him when he left home for the airport, disconcerted her. While acknowledging the airline's right to certain of her husband's things, Anne always insisted that three pages of personal notes associated with the briefing session documents were private property. She had glanced at them while reviewing Greg's flight documents almost three weeks earlier. They were in her house then, along with the briefing folder, yet all her husband's paperwork had disappeared after the liaison officer's arrival on the scene. If he hadn't taken the personal notes, then where did they go?

Assuming that Captain Crosbie *had* removed the several pages of personal notes along with other flight-related documents of her husband, Anne Cassin made multiple requests to Air New Zealand management to return those notes to her. She never got them back. The reason could only be that they confirmed that the accident aircraft's intended route was up McMurdo Sound.

When Justice Mahon's *Verdict on Erebus* came out in 1984, Anne Cassin was shocked to read that the relative who had originally told her that he had handed over material to the liaison officer at *that officer's explicit behest* had subsequently given the airline a letter verifying that, in fact, he had done so *on his own responsibility*. It seems doubtful that Anne could have guessed her brother-in-law's reason for altering a material "fact" regarding on whose initiative her husband's things disappeared from her home without her permission or even knowledge. However, the spirited woman quickly made clear that she was decidedly allergic to being lied to.

First officer Greg Cassin's widow was to have a rough life even after the Erebus disaster had wreaked havoc on it. In 1999, she went blind from a stroke, which would be hard for anyone but close to insupportable for a professional pilot who earned a living in the sky. Nevertheless, within a year Anne was flying again. Since, legally, she was allowed to do so only if there was a qualified instructor on board, she now flew under the supervision of a pilot friend who ran Tasman Bay Aviation. Fate struck again, though, when Anne was diagnosed in her early sixties with Parkinson's disease. She took to calling herself a "gray-haired vision-impaired lady with Parkinson's disease" who loved sailing, "especially hands-on sailing." The woman had guts to spare.

As for the family of Captain Collins, its members for years suffered public disgrace because of the airline's strongly promoted contention that, as the pilot in command, Captain Collins personally was responsible for TE901's collision with an Antarctic mountain. The grace under pressure that Maria Collins consistently exhibited despite this heavy burden is truly extraordinary. It reveals a rare emotional fortitude that we can regard as yet another variation of courage.

Captain Collins's widow and the two eldest of her four daughters politely but firmly disputed several of Chief Inspector

Chippindale's preposterous claims involving them. They also spoke out clearly about Captain's Collins's fastidious preparations just before the aerial tour he would be commanding. This evidence played a powerful role in persuading the dodgy Privy Council to rule that Justice Mahon had indeed been right about the dominant cause of the disaster. It's hard to insist, as the airline executives had interminably been doing, that Captain Collins *knew* that his flight path was routed over Ross Island when he'd spent the night before his trip explaining his track up McMurdo Sound to the girls as he plotted it out on an extra-large topographical map.

In these cases, and others as well, the individuals involved refused to remain silent about what they knew or suspected touching on the accident or its aftermath. Each thereby contributed to our overall understanding of how Air New Zealand and the Muldoon administration that had its back went about eliminating evidence that proved the airline was responsible for the crash, planting evidence that suggested the pilots were responsible for it—or manipulating evidence (like the CVR) if neither elimination nor planting was an option. These courageous individuals all constituted bright spots in a saga marred by gross abuses of power by those most responsible for upholding the rule of law—*and for keeping the flying public safe.*

How do abuses of power originate and proliferate so seemingly easily? What is the thinking of those willing to embrace a strategy of deception? During their deliberations concerning the merits of Justice Mahon's appeal, the Judicial Committee of London's Privy Council briefly contemplated the matter in an exchange between Lords Diplock and Templeton. Although their provisional answer was wrong, it was a terrific question.

22

THE ANSWER TO LORD DIPLOCK'S QUESTION

THE HIGH-CEILINGED, OAK-PANELED CHAMBER AT 9 Downing Street had filled to capacity when the Privy Council's Judicial Committee began its fourteen-day hearing of Peter Thomas Mahon's appeal against the New Zealand appellate court's rulings regarding his findings as royal commissioner into the Erebus disaster. At one point, Lord Diplock inquired when the conspiracy Mahon had uncovered was supposed to have started. Lord Templeton volunteered that, for his part, he found it very hard to believe that Air New Zealand CEO Morrie Davis and key airline executives sat around the day after the crash cooking up a scheme to deceive everyone as to its cause.

When and how *do* cover-ups happen? Typically, they start small and metastasize over time. Occasionally, though, the groundwork *is* laid literally overnight. Either way, cover-ups wind up badly tarnishing the reputations of the very organizations they were designed to protect. All that's needed initially is, as Stuart Macfarlane classically put it, "a public disaster carrying highly embarrassing overtones."

The horrific crash of a sightseeing Air New Zealand jetliner at the bottom of the world was such a precipitating public catastrophe. As too frequently happens with top officers at high-profile companies and even the Catholic Church, airline CEO Morrie Davis reacted instinctively by circling the wagons. For the time being, he declined to say anything about the deadly, last-minute flight plan change of which the pilots, in violation of international aviation safety protocols, had not been informed. Davis didn't tell the Ministry of Transport's chief inspector of air accidents, Ron Chippindale, who was tasked with speeding to the distant crash site and investigating the accident's cause. He didn't tell the airline officers deputized to interpret and transcribe the CVR in the US with the NTSB's expert assistance. He didn't even tell his own board of directors at its first meeting after the accident, which was a full week after it had occurred.

Superficially, then, it might seem that nothing more was at work than Morrie Davis's instinct, shared with legions of other challenged company leaders, to stall for time. That the main thesis of the airline's eventual cover-up story had yet to be developed can be seen from the CEO's retrospectively indiscreet statement at the board meeting. There he had stated that the only clue so far to the tragedy's cause was that the accident aircraft was "left of track," which could only mean left of the 1978–1979 flight track up the center of McMurdo Sound, twenty-seven miles away from the crash site on Ross Island's Mount Erebus. The fictitious storyline that the route always lay over a dangerous (and active) Antarctic volcano, contrary to the strict safety protocols of the Americans who controlled the area's airspace, would come later.

There *had* been one curious development, however, on the very day TE901 disappeared. It was initiated by a phone call at 8:50

p.m. from Captain Eden, Air New Zealand's director of flight operations, to the head of the Ministry of Transport's Office of Air Accidents Investigation, Ron Chippindale. After being advised that the aircraft was missing and probably down, Chippindale swiftly assembled the small party that would accompany him to Antarctica the next day to investigate. One member of his team, he decided, would be the airline's Captain Ian Gemmell.

From the standpoint of a normal air disaster inquiry, appointing a principal target of that inquiry to help with the investigation would be bizarre. As the real architect of the Antarctic flights, Captain Gemmell admittedly knew everything there was to know about them—but since it was *his* operational arrangements that were now to be subjected to intense scrutiny, Chippindale must have had some ulterior motive in involving him. It could have been that Chippindale anticipated that Air New Zealand might prove to be responsible for the catastrophe and prudently decided to invite one of its key figures to assist him. That way, they could all better "protect" the airline.

Leaders who find themselves thrust into extremity like airline CEO Morrie Davis and his like-minded pal Prime Minister Rob Muldoon typically do not consider themselves unethical, not to speak of criminal, but simply practical. What has happened may be regrettable, even tragic, but a leader can and should devise a way out for his institution. For long periods of time, even indefinitely, top executives convince themselves that their actions are reasonable and proper considering the urgent, sensitive situations thrust upon them. In the case of a downed DC-10, for instance, the prospect of a rosy outcome for the national airline would fully justify whatever means might be required to ensure it.

At some level, this may appear quite natural. Those at the top of an organization wish to stay in charge of the unfolding narrative, especially if much financially and reputationally is at stake, as it was with the Erebus disaster. The last thing those in positions of authority want to do is surrender the investigation and resolution of their ticklish issues to law enforcement personnel. Indeed, they may well recruit legal experts of their own to hone whatever the corporate line is going to be and defend them in case of litigation.

There will, of course, be many lower-ranking employees within the challenged organization who know something. The stronger institutional traditions are, the more loyalty it will generate in this class of personnel. It has been said that, prior to TE901's crash, Air New Zealand was a revered institution, more like a church than a company. In such a context, the feeling of institutional loyalty will tend to overpower an individual worker's moral compass, preventing him or her from clearly perceiving organizational wrongdoing.

The desire to retain one's own job provides a powerful incentive not to report anything that seems like it might land the *reporter* in trouble. Dorday and Greenwood, who originally discovered the undisclosed coordinate change, are examples of these two tendencies. While sharing their terrible discovery with their own superiors, they deliberately withheld it from the two air accident investigators sent to Flight Dispatch explicitly to probe for intelligence and documents late in the day TE901 mysteriously disappeared. (The airline later rewarded Greenwood with the Navigation Section's top job.)

Returning to Lord Diplock's question, Morrie Davis's order for the destruction of masses upon masses of airline documents, especially those bearing on the Antarctic charters but many others too, represents an escalation of his effort to conceal information on Air New Zealand's operations from public view. To what degree Davis's shredding order was aided or even directed by the prime minister himself, who would prove to be at the center of the massive conspiracy, we don't know. Future minister of transport Maurice Williamson, who worked for the airline as a corporate planner during this period, witnessed the spectacle with stupefaction. Initially he believed these reams of material were being collected for those charged with investigating the world's fourth-worst air disaster rather than those bent on preventing potentially relevant evidence from coming into their possession.

Destroying incriminating data—whether from the distant crash site, the homes of the accident flight's pilots, or Air New Zealand's corporate headquarters itself—would have been considered necessary but not sufficient to sell an entire nation on the airline's complete lack of responsibility for a catastrophic disaster involving one of its jets and mostly Kiwi victims. Providentially, the government was able to provide an alibi for Air New Zealand via its Ministry of Transport's Office of Air Accidents Investigation. Here is where the cover-up proper got its start.

The Chippindale Report is decidedly not the product of a single governmental functionary working independently to establish the cause or causes of and culpability for a colossal disaster. It was a collaborative effort, directed from the highest levels of the Muldoon administration with extensive input from the airline in the person of Chippindale's close associate Captain Gemmell, to explain away the accident as pilot error. Chief Inspector Ron Chippindale, who knew his place in the overall scheme, was amenable to coaching and received plenty of it. His superiors must have been pleased with the completed report to which he gave his name, for Chief Inspector Chippindale remained at his post in the Ministry of Transport until his retirement years later. He also helped himself to numerous free flights on Air New Zealand until a new boss at the ministry, Maurice Williamson, cut him off in disgust. (It was Williamson who, twenty years after the Mahon Report came out, finally tabled it in Parliament to give it official status at last.)

Public distrust of the government's internal air accident investigator in time motivated New Zealand's prime minister to reach beyond Chippindale (and Gemmell) to designate an "independent" royal commissioner to conduct a second inquiry. Muldoon's choice was a High Court judge who had been recommended for the post by high-profile Auckland lawyer Lloyd Brown. A close friend of Justice Mahon, Brown himself had already been selected to represent the airline during the Royal Commission of Inquiry proceedings and asked whom he'd like to see run the proceedings.

The selection of Justice Mahon was a calculated risk on the part of the Muldoon administration. The prime minister appreciated that he would need a second conclusive finding of pilot error to convincingly absolve Air New Zealand of any culpability for the loss of a DC-10 and all its occupants. It was assumed that the newly selected royal commissioner, who was a conservative, would appreciate his unspoken obligation to produce a report reinforcing the viewpoint of the administration that had just appointed him. Still, those in charge of preserving the orthodoxy of the Chippindale Report were relying on Brown to use his longstanding ties with the newly appointed royal commissioner to "manage" him should that ever prove necessary. In short, the selection of Justice Mahon as royal commissioner was part of the conspirators' plan for influencing the second investigation's course and outcome.

In England, a preliminary accident investigation by a governmental agency normally did not publish its findings if a Royal Commission was in the offing; the original inquirer merely turned his material over to the second one to use as he saw fit. This scenario would have been anathema to Prime Minister Muldoon and the national airline's top executives. Providentially, in New Zealand a vagary of the law enabled the conspirators, despite criticism, to publish the findings of the initial governmental investigator. By seeing to it that the start date of the Royal Commission was delayed by several months, they were able to get the Chippindale Report's thesis widely disseminated throughout the country before the second investigation could even start, much less conceivably arrive at a wholly different interpretation of events.

Once the royal commissioner commenced his own inquiry, the conspirators faced a new challenge. Key members would have to give evidence, under oath, before Justice Peter Mahon and the able counsel assisting him. What would be their script for that ordeal? The key points appeared in a "smoking gun" document that surfaced only after Justice Mahon's investigation had been completed. The document consisted of the minutes of an Air New Zealand board of directors' meeting shortly before the Royal Commission hearings commenced. Those minutes revealed that the storyline would be (a) the route was always programmed to overfly Mount Erebus and (b) because of that, pilots had to maintain an altitude of sixteen thousand feet until on the volcano's south side. They also revealed that the airline would rely on Chief Inspector Chippindale's view on the accident's cause.

But the route in 1978 and 1979 (except for the accident flight) lay not over Erebus but up perfectly flat McMurdo Sound. Moreover, even in 1977, when the flight path did ostensibly lead over the volcano, no pilots ever overflew it. Even Captain Gemmell, who designed this dangerous track and commanded the initial Antarctic flight, did not claim at the Royal Commission hearings that he had risked going over the top of an active volcano constantly releasing toxic gases into the atmosphere above it. Nonetheless, he defended this track as being superior to flying up McMurdo Sound, making no mention of the fact that McMurdo Station's American air traffic control officers required aircraft approaches to the area to be via the sound.

Easily refutable as it was, the conspirators needed to maintain the fiction that the route was *always* over Mount Erebus because it made possible a second fiction—that all flights had to maintain an altitude of sixteen thousand feet until over the mountain because, otherwise, they would fly into it just like Captain Collins and first officer Cassin had done. This was a highly injudicious stance to take. The 1977 pilots, whose preprogrammed flight path *was* over Mount Erebus, had all been briefed that they were at liberty to disengage the nav track at Cape Hallett and fly visually during the last leg of the outbound journey. It was easy to do by simply swooping up McMurdo Sound parallel to the nearby coastline of Victoria Land and letting local air traffic control clear the aircraft for the final turn eastward to Ross Island's Hut Point Peninsula. As for the 1978–1979 pilots, their flight path had been shifted—in every case but the last—to proceed up the middle of McMurdo Sound to a final waypoint in the Dailey Islands. This put these aircrews well out of the way of the mountain whether flying visually or on nav track after they reached Cape Hallett.

Additionally, to accommodate negative passenger feedback, Air New Zealand had quickly relaxed *in practice* its regulator's formal rule concerning the minimum safe altitude, which was just for overflying the volcano that no one was electing to overfly anyway. From the third of the initial *1977* series of Antarctic charter flights, low flying up the flat sea ice of McMurdo Sound was widely and enthusiastically being reported throughout the country in newspapers as well as in Air New Zealand's own internal newsletters. Written statements by top industry executives who'd enjoyed such a flight were quoted far and wide. Try as they might, the airline's top management had difficulty explaining away this copious, *published* body of evidence confirming the popularity of its jetliners' practice of low flying up McMurdo Sound. Since the route was over flat sea ice forty miles wide east to west, it was perfectly safe to fly within one thousand to two thousand feet of it to give passengers photo ops.

Being one of Air New Zealand's most risk-averse pilots, Captain Collins deliberately did not disengage his aircraft's nav track when he overflew Victoria Land's Cape Hallett, as the other Antarctic pilots appear to have decided in advance to do. That his was a deliberate plan to fly visually, but to *stay* locked on the nav track after the penultimate waypoint is clear from the fact that he took the trouble the night before his flight to plot it onto a topographical chart. Given his reputation for immense care, Captain Collins would have counted on this safety precaution to ensure that he would wind up where he was supposed to at the end of his outbound flight. That did not happen, however, as the airline had reprogrammed that track to take him to a different destination waypoint. Worse, in defiance of aviation safety protocols the world over, no one at Air New Zealand notified Captain Collins of the last-minute switch to his route.

These two missteps, which accounted for the Mahon Report's finding on the dominant cause of TE901's fiery end, were too glaring to be papered over. The leading lights of the national cover-up tried their best to twist or even dismiss the uncooperative facts. Chief Inspector Chippindale, for instance, could find no evidence that the pilots had been deceived about where their aircraft was headed. That it had crashed right when and where it did was not the least bit suggestive to him.

It was the former attorney general of that earlier time, Jim McLay, who would at long last state the obvious. On the fortieth anniversary of what is still the Southern Hemisphere's most deadly civilian air disaster, McLay observed that "Mahon's logic was impeccable. He basically found the plane had been programmed to fly into a mountain, and it did." He added, "You can't really escape that. Mahon got it right, and his finding should be considered the definitive one."

That single finding—let it be crystal clear—contains two logically entwined parts, the second one of which is rarely, if ever, mentioned aloud. The first part is that inexcusably slipshod airline procedures were responsible for the accident. Mahon crisply enumerated them, one after another, in a report that became an international tour de force by virtue of its pioneering contributions to what would become known as systems analysis.

It followed from the judge's causal finding that the Air New Zealand executives themselves were solely responsible for the horrific disaster. Yet in their testimony before the upright judge they had insisted, *for months*, that the pilots alone were responsible. They presented their case to the royal commissioner in an elaboration of their original cover-up story that was, frankly, so preposterous that the provoked judge ultimately exposed it as "an orchestrated litany of lies." Clearly, if it be accepted that the airline's unsafe practices caused TE901's crash, the airline itself *was* responsible for the massive loss of life on Erebus's flanks. And if *Air New Zealand* was indeed responsible, then its top executives' aggressive campaign to attribute blame elsewhere—initially via the Chippindale Report and later via their fraudulent testimony before the royal commissioner himself—could have been *nothing other than* a cover-up.

It would be for other legal minds to pass on the acceptability of a royal commissioner's making a finding of concerted lying by company moguls intent on concealing their own responsibility for a disaster by attributing it to innocent parties. After all, New Zealand had never had to grapple with a multitentacled, national cover-up before. It is repugnant that even now, decades after the accident, influential reporters insist that the blame must be shared between the crew and the airline. Why? Mahon had long since convincingly proved that the dominant cause lay with the airline. Meanwhile, braving Morrie Davis's wrath, Captain Gordon Vette had long since authoritatively discovered that the accident's proximate cause was optical deception, a common phenomenon in polar regions about which Air New Zealand had never briefed its Antarctic pilots. It was sector whiteout, specifically, that had prevented Captain Collins and first officer Cassin from perceiving, *even in clear air,* that they were about to fly into a volcano that they believed (from fixing their position visually and remaining locked on their nav track) was not nearby.

The Muldoon administration's all-out effort to protect the government's bottom line, New Zealand's beloved national airline, and the prime minister's crony who ran it has cast a long shadow over the country's legal and judicial spheres. It has as well rent the social fabric. Especially unfortunate in the Erebus context was Muldoon's characteristically abrasive, willful, and intimidating mien. Adept at operating the levers of power, this alpha male had no compunctions about publicly humiliating and punishing critics, especially one who would take the high moral ground in an accident investigation report. In his cynical view, sacrificing the stellar reputations of two dead pilots and a High Court judge was a relatively small price to pay for safeguarding the airline's legal, financial, and reputational interests, which in this messy case were identical with those of Prime Minister Muldoon's administration.

Confusion and even bitterness have been the legacy of the strife caused by the unprecedented crash in Antarctica of a sightseeing jetliner and the tragedy's highly litigious, protracted aftermath. There are practical ramifications to the current unsettled situation too. For example, it is in the public's interest to know whether the airline ever undertook the overhaul of its dysfunctional operations that Justice Mahon had called for in the interests of aviation safety. In his official report, he'd offered a long list of urgently needed improvements. How many were instituted? It is not clear that anyone knows.

23

APOLOGY

"IN A CENTURY OR TWO, THERE WILL BE AN APOLOGY from the airline, but it's too painful to do in our lifetime." Such were the sentiments of Captain Jim Collins's widow, Maria Collins, on the morning of November 28, 2019. At a gathering later that day for relatives of the 257 individuals instantly annihilated four decades earlier when a sightseeing Air New Zealand DC-10 slammed into an Antarctic volcano, she would be in for a big surprise.

"The time has come," Prime Minister Jacinda Ardern announced, "to apologize for the actions of an airline then in full state ownership, which ultimately caused the loss of the aircraft." Her apology was "wholehearted and wide-reaching," Ardern declared. She had read the Mahon Report, the statements made when it was tabled in Parliament in 1999, and the Privy Council's report after becoming involved in the project to build a national Erebus memorial. Why was there no national memorial already? she had wondered. "It made no sense to me."

After studying these sources, the prime minister continued, it "all built a picture of unfinished business, and it wasn't right." She reasoned that if the government accepted the Mahon Report—"and *we do*"—then an apology was long overdue. The Privy Council's Law Lords, she reminded her audience, had found the royal commissioner's *substantive* findings—which absolved the pilots of any culpability for the terrible crash—totally compelling. New Zealand's own historical resistance to that finding, she speculated, was a "hangover from Chippindale." In an allusion to Justice Mahon's detailed exposition of systemic failures in the airline's organizational arrangements themselves, Ardern also reminded those attending the gathering of his "huge contribution" to how aviation accidents around the world are investigated today.

For completeness, Prime Minister Ardern felt that Air New Zealand ought also to issue a formal apology both for the crash and for its highly insensitive treatment of the bereaved families afterward. There was, she noted, no resistance at the airline to her proposal. Air New Zealand's board chairperson, Dame Therese Walsh, thus obliged her during that same fortieth anniversary gathering.

Ardern's masterful attempt to complete an important piece of her country's unfinished Erebus business seems to have been extremely well received by the accident victims' families. Maria Collins was asked why she thought the youthful prime minister, who was not even born when the air disaster had occurred, could manage to say "I'm sorry" when no earlier administration could bring itself to do so. "It's easier to apologize when you've had nothing whatsoever to do with it," the insightful widow observed. There's "not a hint of guilt, and you can blandly forgive the sins of the forefathers, as it were, because you had nothing to do with it." As for Justice Mahon's widow, Margarita Mahon, she had not attended the November 28, 2019, gathering because she had always felt Erebus anniversaries belonged to the families of the Erebus victims. Instead, she got the news at home on her iPad. Asked later what her husband's reaction to it would have been, Margarita mischievously replied that "[Peter] would have thought the apology was gracious . . . but forty years too late?"

———— ◈ ————

How gracious was the apology, though, in truth? One element critical to resolving Erebus's still unsettled legacy had not even been broached. Did or did not Air New Zealand, which the prime minister had just assured the public *was* the cause of New Zealand's worst civil disaster, conspire to cover up its responsibility for TE901's crash? Blithely to pass off a deviously elaborate, multipronged cover-up as a "hangover from Chippindale" grossly distorts the trajectory of the Erebus story. The Chippindale Report, Ardern implied, simply got it wrong, which misled the country for forty years.

Why did not Ardern follow up her passionate assertion that Justice Mahon had been correct about what caused the Erebus tragedy with a frank admission that the Chippindale Report, far from being the product of a single inept government employee, was in fact a well-planned team effort to delude the public? The report relied very heavily on the input of the imperiled Captain Gemmell, who had everything to lose if the cover-up was exposed. As the architect of the airline's Antarctic charters, he was technically capable of steering the less knowledgeable Chippindale (and the airline's executives) on how to hide Air New Zealand's culpability for the crash at the bottom of the world—and he took full advantage of all opportunities to do so.

In her fortieth anniversary remarks, it appears that the prime minister either wished to avoid broaching the culpability matter or did not appreciate that she *should* have broached it. Her international renown for honesty, empathy, and transparency suggests that Ardern would not have deliberately ignored this thorny topic if she had understood that it existed. The Privy Council had, however, deliberately screened it from public view by claiming that while some airline executive witnesses had undoubtedly lied individually to the royal commissioner, there had been no concerted lying and hence no conspiratorial effort to shift blame for the disaster from themselves to the dead pilots. If Prime Minister Ardern had grasped that the royal commissioner was right about *both* cause and culpability, she would surely have concluded her moving fortieth anniversary speech differently—with a sincere apology to the restless spirit of the Honorable Peter Thomas Mahon, who was more honorable than most. He discovered the grave miscarriage of justice being committed before his eyes and was too honest himself not to call it out.

New Zealand continues to have unfinished business. Meanwhile, halfway around the world, in the United States, a new aviation industry scandal has been unfolding over a period of years. This one, like the earlier one in New Zealand, involves a once-hallowed company whose complacency about safety combined with an indulgent regulator's predisposition to rubber-stamp even its most questionable initiatives to produce two catastrophic air disasters in quick succession. Suggestively, Boeing's CEO, Dennis Muilenberg, who was at the helm when the 737 Max 8 jets went down in Indonesia and Ethiopia, proved every bit as heavy-handed as Air New Zealand CEO Morrie Davis (and the prime minister who backed him) in launching a campaign of misinformation and sophistry intended to clear his company of responsibility for 346 needless deaths. Particularly odious were Muilenberg's efforts to intimidate the grieving victims' families, which was a tactic not unknown to Davis.

When Peter Robison's *Flying Blind: The 737 Max Tragedy and the Fall of Boeing* debuted late in 2021, it was celebrated as a highly readable, if ultimately depressing, case study of business misman-agement. The aircraft manufacturer seems to be getting away with its extensive malfeasance, in the sense that it continues to prioritize its bottom line over aviation safety and has expressed no contrition over having willfully created a bad plane on the cheap that swiftly killed 346 people in two countries. Boeing has managed to survive anyway as a too-big-too-fail shadow of its former esteemed self.

The creators of a cover-up seem constitutionally unable to come clean, even years after they've been found out and publicly exposed by, say, a shrewd royal commissioner able to stitch together the various pieces of their grand deception. Retired Air New Zealand executive pilot Peter Grundy, for instance, took the opportunity of Erebus's fortieth anniversary to remind his compatriots that the pilots alone caused the tragedy. In an oblique reference to the dwin-dling number of conspirators still alive, he remarked that there were "only three of us left now" to explain the accident's cause correctly. It was Captain Collins's decision to descend below minimum safe altitude (in overflying an active volcano that no Air New Zealand pilots, including him, ever went over) without knowing precisely where he was because he never fixed his position (he *had* fixed his position—he was just wrong about it owing to ocular deception).

Whether Jacinda Ardern's successor as prime minister will complete her laudable project to move New Zealand beyond the Erebus tragedy and its explosive, protracted aftermath remains to be seen. If she never learns that the courageous royal commissioner who established the accident's cause *also* discovered an extensive, aggressive, and highly coordinated cover-up of the airline's culpa-bility run from the very seat of government itself in the person of an earlier prime minister, it will suggest that Maria Collins wasn't far off in her thinking. Perhaps a *complete* Erebus reckoning won't be done in our lifetime.

Maybe in a century or two.

PHOTO
GALLERY

THE DC-10'S WRECKAGE AS SEEN FROM ABOVE

Photograph by Nigel Roberts for the Antarctic Division of New Zealand's Department of Scientific and Industrial Research
This photo of the crash site was taken from a helicopter on November 29, 1979, the day after TE901's catastrophic implosion.

COMPANY LOGO ON THE AIRCRAFT'S BROKEN TAILPIECE

Photograph by Nigel Roberts for the Antarctic Division of New Zealand's Department of Scientific and Industrial Research
This iconic shot quickly came to be regarded as a symbol of the tragedy.

BODY RECOVERY WORKERS SETTING UP CAMP ON THE ICE

Photograph by Colin Monteath for the Antarctic Division of New Zealand's Department of Scientific and Industrial Research

The men's daunting mission was to chip out bodies and body parts for transport back to New Zealand for identification and return to the victims' loved ones.

Tents Uncomfortably Close to the Work Site

Photograph by Colin Fink
The work area extended from the DC-10's point of impact at the bottom of the sloping crash site all the way up to the end of the wreckage. Recovery workers were assigned a grid and told to remain there until all human remains found in it had been processed.

A Scene of Devastation for Man and Machine

Photograph by Colin Fink
The huge pressures and counter-pressures created when 250 tons of aircraft (and its contents) hit Erebus's rock-hard, ice-covered slopes triggered titanic shock waves, the initial one of which instantly killed all on board TE901.

THE TERRA NOVA MEMORIAL CROSS, A
HISTORIC ANTARCTIC MONUMENT

Photograph by Nigel Roberts
Situated on the summit of McMurdo's Observation Hill, the nine-foot cross honors Robert Falcon
Scott's polar party of five that reached the South Pole on January 17, 1912 – five weeks after their
Norwegian competitors led by Roald Amundsen.

The Cross's Inscription

Photograph by Nigel Roberts

The inscription consists of the final line of the closing passage of Tennyson's "Ulysses," a paean to new experience. Here you can barely make out the words *To strive, to seek, to find, and not to yield*. Scott's group died just eleven miles short of a supply depot on the Ross Ice Shelf, which the cross faces.

ACKNOWLEDGMENTS

WITHOUT THE AUTHORITATIVE INPUT OF MULTIPLE specialists on one or another facet of Erebus, I could not have produced a work anywhere near as factually accurate, multidimensional, and informative as this one aspires to be. My outreach to New Zealand began with Sir David Baragwanath, an urbane jurist of legendary fame who long ago served as senior counsel assisting the Royal Commission. Later, I made contact as well with retired judge Gary Harrison, who simultaneously served as junior counsel and, sometime later, as legal adviser for *Erebus, The Aftermath*. The latter shared with me the biggest bombshell I can safely reveal in this book, valuable reminiscences of his Royal Commission work, and his perspective today on the conflagration that erupted when the Mahon Report's findings proved to be the antithesis of those in the Chippindale Report.

Successful American aviation journalist and author Christine Negroni was an early inspiration to me even though, after solicitously reading the original beginning of my manuscript, she suggested I scrap it. It was very good advice, however, so I wrote a new, more intriguing introduction to this unusual tale of an inexplicable air accident made notorious by a well-coordinated plan by those responsible for it to cover their tracks. *Judgment on Erebus* thus now opens in medias res, plunging readers into the sensational saga the evening before all hell breaks loose across New Zealand.

Judge Peter Spiller, a former distinguished academic with a profound understanding of the Court of Appeal and the Privy Council, kindly took me under his wing and, over several years, brought me up to speed about how these legal bodies worked in that era. He also patiently critiqued the "legal chapters" of my draft manuscript, thereby sparing me much embarrassment. A supremely gracious mentor, Judge Spiller encouraged my ambition to reassess Justice Mahon's evidence in support of findings that, once made public, Prime Minister Muldoon and Air New Zealand executives summarily denounced as unbelievable.

The guardian angel of Erebus researchers is Stuart Macfarlane, who goes by no official designation. Prior to the crash of TE901, he was first an attorney and later a law lecturer at Auckland University. During the years since the jetliner's loss, Macfarlane has mastered every facet of the horrific accident and it's even more horrific aftermath, turning himself in the process into an oracle. There are no words to describe the electrifying effect this guardian angel of Erebus researchers has had on me. His zest for helping others grasp just how devious an assault was mounted on New Zealand's legal and judicial systems in the wake of the world's fourth-worst air disaster clarified multiple murky points for me. Going above and beyond in his helpfulness, Stuart Macfarlane graciously agreed to read my entire manuscript, along with all my notes. I was prepared for a drubbing and was deeply relieved when one was not forthcoming.

Through the kind offices of Ms. Kathryn Collins Carter, Captain Jim Collins's eldest daughter, I was introduced to the remarkable Reverend Dr. Richard Waugh, who combines pastoral duties with writing mesmerizing books on the history of air disasters and aviation safety in New Zealand's pioneering era of small planes (~1945–1970). Of course, he is no slouch when it comes to a big aircraft disaster like that of TE901, and we have enjoyed a long and productive correspondence regarding it. Serving as a second technically informed reader of my entire manuscript, Rev. Waugh offered advice on word choice, phraseology, and much else. Among his pastoral specialties is ministering to the families of air disaster victims and ensuring that memorials to them get erected. As the "instigator" (his word) in 2016 of the Erebus National Memorial, Rev. Waugh is making steady progress toward achieving it despite multiple unanticipated obstacles.

To Dr. Nigel Roberts, emeritus professor of political science at Victoria University of Wellington, I owe a great debt of gratitude for sharing the iconic photographs he took of TE901's wreckage the day after the disaster. (He was then a young information officer and photographer stationed at Scott Base, a New Zealand scientific research facility in Antarctica.) Since "sharing the photographs" proved much more challenging than envisioned, I wish here publicly to thank Dr. Roberts for sticking with the project despite the hassles.

I wish also to thank Dr. Grant Morris of Victoria University of Wellington's law faculty for his contributions to this work through both his own research on Justice Mahon and our correspondence, which included (on his end) good-naturedly answering questions as well as critiquing a particular chapter of my manuscript. "I teach [Erebus] in my 'famous cases' module," noted Dr. Morris, who considers Mahon "an example of someone in the elite that takes a stand against powerful people on the basis of conscience." This trenchant observation accounts for the "conscience" in *Judgment on Erebus*'s subtitle.

I cannot fail to express my appreciation to Dr. James Jose Bonner, a retired developmental biologist with wide-ranging intellectual interests and an enviable facility with computers. It is he who read a very early draft of my manuscript, indicated where it was thematically weak, and volunteered to produce all my work's maps. It is not every Erebus researcher who is blessed with such a capable and big-hearted brother.

Last but not least, I wish to acknowledge my gratitude to Al. Gibbons, who does not share his wife's passion for air disasters and aviation safety, for civilly enduring nearly a decade of my either researching Erebus or writing about Erebus when not engaged in my entrepreneurial day job. Al. has also improved the book by making valuable strategic contributions to it from time to time.

SELECT GOVERNMENT-CONNECTED DRAMATIS PERSONAE

LLOYD BROWN, a celebrated Auckland attorney and lead counsel representing Air New Zealand during Royal Commission hearings presided over by a judge he himself had commended to the Muldoon administration for the post

RON CHIPPINDALE, head of the Ministry of Transport's Office of Air Accidents Investigation and author of the government's in-house official report on TE901's crash

BRUCE CROSBIE, Air New Zealand pilot involved in the disappearance of two pieces of crucial evidence that would have disproved the veracity of the Chippindale Report's findings

MORRIE DAVIS, the tough, autocratic CEO of Air New Zealand

DES DALGETY, Muldoon's private attorney and man on the board of Air New Zealand, responsible for Air New Zealand's engagement of Lloyd Brown and thought by Mahon to be the orchestrator of the litany of lies

ALAN DORDAY and *DAVID GREENWOOD*, both of Air New Zealand's Flight Dispatch Unit, who discovered the computer programming error in Captain Collins's flight plan but did not report it to investigators seeking information about the missing plane

IAN GEMMELL, Air New Zealand's chief pilot and flight manager (technical), who designed the original 1977 flight path over Mount Erebus and was a crucial figure in the attempted cover-up of Air New Zealand's responsibility for TE901's crash into Mount Erebus

DAVID GREENWOOD—see under ALAN DORDAY above

BRIAN HEWITT, Air New Zealand's chief navigator, who in 1978 unwittingly reprogrammed the 1977 flight path from over Mount Erebus to over McMurdo Sound's flat sea ice, then in 1979 unwittingly reprogrammed the flight path back to over Mount Erebus

ROSS JOHNSON—see under JOHN WILSON below

EDGAR KIPPENBERGER, director of the Ministry of Transport's Civil Aviation Division (CAD, Civil Aviation), which was Air New Zealand's lenient regulator

ROB MULDOON, New Zealand's imperious prime minister (1975–1984), who was Air New Zealand's shareholder representing the government

GORDON VETTE, Air New Zealand's top pilot, who did not accept the government's in-house analysis of where the air disaster's cause and culpability lay. Assisted Justice Mahon in his quest to determine the causes of TE901's crash. Researched optical illusions to determine how an outstanding crew could be fatally deceived into smashing into a volcano in broad daylight. Unceremoniously cashiered by Air New Zealand for not supporting the pilot error explanation of the accident.

JOHN WILSON, supervisor of Air New Zealand's Route Clearance Unit, and *ROSS JOHNSON*, flight manager (line operations), who briefed five pilots (including Captain Collins, first officer Cassin, and Captain Simpson) on their impending Antarctic flights

GLOSSARY

AREA INERTIAL NAVIGATION SYSTEM (AINS)
The then state-of-the-art onboard electronic navigation system that, deploying three inertial sensors, enabled aircraft to be carried *unerringly* along a preprogrammed computerized flight path for thousands of miles

AIR TRAFFIC CONTROL (ATC)
Ground-based controllers of a designated air space that monitor, give instructions, and advise pilots entering or leaving the area over which they have jurisdiction

> **MAC CENTER:** The Americans' principal complex at McMurdo Station (for high-frequency radio communications)

> **ICE TOWER:** The Americans' ancillary installation near Williams Field's ice runway (for radar and for very high frequency radio communications, both of which required line-of-sight access)

"AN ORCHESTRATED LITANY OF LIES"
One of New Zealand's most well-known quotes, intended by Justice Mahon as a searing indictment of a massive, multitentacled, government-wide attempted cover-up of the national

airline's responsibility for an infamous crash by aggressively pinning the blame on the innocent dead pilots

AZIMUTH
An angular measurement in a spherical coordinate system

BEEHIVE
The executive wing of New Zealand Parliament Buildings in Wellington

COORDINATE SYSTEM (LATITUDE AND LONGITUDE)
A means by which to locate any place on Earth's surface, with latitude being a measurement of its position north or south of the Equator and longitude being a measurement of its position east or west of the prime meridian at Greenwich

DISASTER VICTIM IDENTIFICATION (DVI) SQUAD
New Zealand policemen sent to the hazardous crash site to "locate, tag, photograph, and bag" bodies for return to their grieving relatives at home

DIVINE COMEDY, THE
The fourteenth-century epic poem by Italy's Dante Alighieri, much studied and admired by Justice Mahon

FLIGHT RECORDING DEVICES (formerly known as "black boxes," an outdated term since today they are bright orange)

> **Cockpit voice recorder (CVR):** Electronic device that records the most recent half hour of sounds in the cockpit, including the aircrew's dialogue

Flight data recorder (FDR): Electronic device that monitors the dozens of instructions constantly sent to an aircraft's electronic systems

GO-AROUND POWER
Extra power that a pilot applies when alerted that a problem may exist—different from the maximum thrust used when a pilot perceives a dire emergency

> **Origin of term:** Instances in which a pilot fears not making a clean landing and decides to pull up and go around the airport to try again

GROUND PROXIMITY WARNING DEVICE (GPWD)
Safety feature that, at the time of TE901's crash, gave only six seconds' warning

HORIZONTAL SITUATION INDICATOR (HSI)
Instrument on the flight instrument panel that shows the distance to the next waypoint in the top left-hand corner

INTERNATIONAL CIVIL AVIATION ORGANIZATION (ICAO)
An important global body with many signatories, including New Zealand and the United States, with Captain Sully Sullenberger currently serving as the American representative

LANDMARK
An identifiable topographical feature

MAN FOR ALL SEASONS
Label used historically to describe Sir Thomas More and appropriated by Sam Mahon to describe his father as well since both were individuals who acted out of conscience

MINIMUM SAFE ALTITUDE (MSA)
A generic expression denoting an altitude below which it is unsafe to descend, typically because of terrain

MOUNT EREBUS
The southernmost active volcano on Earth and the second-tallest volcano in Antarctica, located on Ross Island in the slice of the continent directly south of New Zealand

NATIONAL TRANSPORTATION SAFETY BOARD (NTSB)
Universally admired American institution under whose auspices the original transcription of the cockpit voice recorder was made by carefully selected New Zealand experts who knew the crew as well as the aircraft

NATURAL JUSTICE
A legal term about whose requirements Justice Mahon and the Court of Appeal's five members (and also the Privy Council) differed

NAVIGATION TRACK (NAV TRACK)
The flight path carried on board an aircraft in its computer

OPS FLASH
A line on the flight plan given to pilots just before boarding that indicates whether there have been any last-minute changes to it

PILOT ERROR
Along with mechanical failure, a traditional causal category used by air accident investigators before Justice Mahon and Captain Vette pioneered the creation of a new causal category: organizational disaster

PILOTS, TYPES OF

Executive pilots: Pilots that combine flying with administrative duties (hence the saying, "to fly a desk")

Line pilots: Pilots that exclusively fly airplanes

PRIVY COUNCIL (in London)
The highest court of appeal under New Zealand's legal system at the time of TE901's crash, with cases reviewed by its Judicial Committee

VISUAL FLIGHT RULES (VFR)
The rules pilots must observe when flying visually

VISUAL METEOROLOGICAL CONDITIONS (VCM)
Conditions in which it is safe for pilots to fly visually

WAYPOINTS
The latitude-longitude means of specifying an aircraft's exact flight path, with "destination waypoint" indicating the last in a series

WHITEOUT
Optical illusions of multiple sorts, one of which—sector white-out (a.k.a. "flat-light" illusion)—involves having excellent vision in three compass directions but not the fourth

WOODHOUSE JUDGMENT
The Court of Appeal's splenetic minority judgment, which Justice Mahon said caused him to resign from the bench

NOTES

NOTES TO PROLOGUE

1. **Justice Peter Thomas Mahon QC** The lifetime rank of Queen's Counsel (or King's Counsel) is an honor established in New Zealand in 1907. The conferment of QC usually requires an extended period of distinguished practice at the bar. Mahon had "taken silk" in 1971, the same year he was appointed a High Court judge.

2. **"Tomorrow all hell's going to break loose."** Sam Mahon recounts his father's visit in *My Father's Shadow: A Portrait of Justice Peter Mahon* (New Zealand: Longacre Press, 2008; repr. 2009; hereafter cited as MFS), Kindle. See MFS, pt. 1, "Lago Como, Italy."

3. **convivial luncheon with his close friend Lloyd Brown** James McNeish, *Breaking Ranks: Three Interrupted Lives* (New Zealand: HarperCollins, 2017; hereafter cited as BR), p. 226.

4. **Muldoon was particularly open to the advice of his personal attorney** McNeish, BR, p. 226. By virtue of his dual positions as the prime minister's personal attorney and his man on Air New Zealand's board, Dalgety played an outsized role in the extended Erebus affair. "There are a lot of untold stories there," observed Marilyn Waring in her foreword to Nicky Hager's *The Hollow Men: A Study in the Politics of Deception* (New Zealand: Craig Potton Publishing, 2006).

5. **described as "a good conservative chap"** McNeish, BR, p. 227.

6. **the Honorable Peter Thomas Mahon** Judges of all superior courts are referred to formally by the style "the Honorable," both while holding office and afterward.

7. **There was an air of the outsider** John Burn, in *White Silence*, ep. 4, "The Love Song of J. Alfred Prufrock," November 10, 2019, podcast, produced by Stuff and RNZ, cohosted by Michael Wright and Katy Gosset, https://shorthand.radionz. co.nz/white-silence/.

8. **raised as a Catholic by a devout Irish grandmother** Margarita Mahon, *White Silence*, ep. 4, "Love Song." See also McNeish, BR, p. 214.

9. **"a sense of 'being different' was always there"** Janet Mahon, quoted in McNeish, BR, p. 214.

10. **reserved, complicated individual of few words** See Margarita Mahon in *White Silence*, ep. 4, "Love Song." What Margarita calls her "more or less proposal" consisted of a piece of paper on which was written T. S. Eliot's "The Love Song of J. Alfred Prufrock." The poem opens with a quotation from Dante, one of Peter Mahon's favorite authors, that is meant to suggest that Prufrock, like Count Guido da Montefeltro, is in hell—albeit a hell on earth. In his tortured internal monologue, the timid, insecure Prufrock wrestles with whether to approach any of the women at a social gathering but fearfully postpones doing so on any number of neurotic grounds.

After she'd read the poem, the diffident suitor asked Margarita what she thought of it. While her exact reply is not known, she has confessed that, in her long relationship with Peter, she had to guess all the time.

11. **"admired people who sort of stuck their neck out"** Burn quoted in *White Silence,* ep. 4, "Love Song."

12. **champion of "fairness and justice," "ready for a joust with authority," "high-handed actions"** Colin R. Pidgeon, quoted in Stuart Macfarlane, *The Erebus Papers: Edited Extracts from the Erebus Proceedings with Commentary* (Auckland: Avon Press, 1991; hereafter cited as EP), pp. 6, 5.

13. **According to Sam Mahon, his father was doing most of the work** MFS, pt. 2, "Album," "Murder."

14. **"Peter Mahon and I sat up all night," "went through everything we could find"** Brian McClelland, quoted in *White Silence,* ep. 4, "Love Song."

15. **"feel obliged to seek leave to withdraw from the case"** P. Mahon, quoted in S. Mahon, MFS, pt. 2, "Album," "Murder."

NOTES TO CHAPTER 1:
THE ALLURE OF ANTARCTICA

1. **"emitting flame and smoke in great profusion," "great height"** James Clark Ross, quoted in Apsley Cherry-Garrard, *The Worst Journey in the World* (1922; repr., New York: Skyhorse Publishing, 2016), p. 10.

2. **warships of the bomb vessel type** For an outstanding description of James Clark Ross's expedition, see Michael Palin, *Erebus: One Ship, Two Epic Voyages, and the Greatest Naval Mystery of All Time* (Vancouver: Greystone Books, 2018), pp. 35–185. So sturdy and reliable did *Erebus* and *Terror* prove on Ross's Antarctic expedition that the two vessels would later be used for Sir John Franklin's ill-fated 1845 Arctic expedition.

3. **Race to the South Pole** For a lively account, see Hunter Stewart, *South: Scott and Amundsen's Race to the Pole* (self-pub., CreateSpace, 2017).

4. **"The worst has happened," "Great God! This is an awful place"** Robert Falcon Scott, quoted in David Day, *Antarctica: A Biography* (New York: Oxford University Press, 2013), p. 148.

5. **final line of the closing passage of "Ulysses"** That action is heroic and preferable to inaction, whatever the ultimate outcome, reaches back to Dante and, beyond him, to Homer for inspiration. *The Divine Comedy's* apocryphal account of Ulysses/Odysseus's return home from the extended ordeal

of the Trojan War makes the message crystal clear. Instead of luxuriating in his reunion with his long-suffering wife, Penelope, he is bored and sets off on a new seafaring adventure, which proves fatal.

6. **Operating out of New Zealand and Australia** An excellent video illustrating how majestic the views were, and remain, from aircraft touring over Antarctica at 16,000 or 18,000 feet is on YouTube; see DennisBunnik Travels, "Qantas 787 over Antarctica—the Most Beautiful Flight Review Ever!" April 11, 2021, YouTube video, https://www.youtube.com/watch?v=T0IfIeAKcgk.

   This 2021 Qantas flight took a route directly south of New Zealand, which put the aircraft at one point in the same slice of West Antarctica that the Air New Zealand charters overflew during their heyday (among the features shown is Mount Erebus).

7. **A supremely conscientious and technically proficient pilot** Captain Collins's integrity and skills appear to have been universally admired. Even Air New Zealand's formidable CEO Morrie Davis described him as "probably one of the most outstanding pilots in the airline"; quoted in Gordon Vette and John Macdonald, *Impact Erebus* (New York: Sheridan House, 1983; repr. 1984; hereafter cited as IE), p. 164.

8. **lightning bolt blew off two compartment doors** The terrifying incident that occurred on Captain Collins's flight out of LA is described in Vette and Macdonald, IE, p. 109.

9. **"As long as you have wings and engines"** Mathew Dearnaley, "One Moment, and 25 Years of Pain," *New Zealand Herald*, June 2, 2005, https://www.nzherald.co.nz/nz/one-moment-and-25-years-of-pain/7TJUQLEAD2OQ6HR3BAYF2L RDG4/.

10. **"as skilled, experienced and dedicated as any"** Vette and Macdonald, IE, p. 25. As members of a relatively small airline, Air New Zealand pilots and flight engineers flew together often enough to know "each other's mannerisms and inflections," thereby enhancing intra-crew communication on a flight; Vette and Macdonald, IE, p. 141.

11. **three mesmerizing earlier trips over the North Pole** Vette and Macdonald, IE, p. 82.

12. **"as if touched with white fire"** Peter Mahon, *Verdict on Erebus* (Auckland: William Collins Publishers, 1984; hereafter cited as VOE), p. 193.

13. **attended a company briefing** The other three attendees at the November 9 briefing were Captain Simpson, first officer (relief captain) Gabriel, and first officer Irvine, who were scheduled to man the second of November 1979's four scenic tours. The briefing is described in Vette and Macdonald, IE, pp. 95–102.

14. **condition attached to Air New Zealand's Air Service Certificate No. 22** P. Mahon, VOE, p. 103.

15. **carrier quickly and quietly did away with this safety feature** The three military services operating in this sector of Antarctica, in contrast, required pilots to have a minimum of

three previous flights into the region under the supervision of an experienced polar flier; Vette and Macdonald, IE, p. 100.

16. **In a further safety lapse** P. Mahon, VOE, pp. 115–116.

17. **could be no dispute regarding the flight's outbound terminus** Captain Simpson at the Royal Commission hearings; see the relevant testimony cited in Macfarlane, EP, pp. 681–682.

18. **Kathryn and Elizabeth, the two oldest of his four daughters** P. Mahon, VOE, pp. 215–216.

19. **"well below the towering volcano Erebus belching smoke"** *Auckland Star,* October 22, 1977, quoted in P. Mahon, VOE, p. 76.

20. **passed over McMurdo Station and Scott Base** The Americans' McMurdo Station was established in 1955 as part of American initiatives collectively known as Operation Deep Freeze. Today, McMurdo Station is the largest base in Antarctica. In 1979 the US Navy was still handling logistics for the civilian program, which is run by the National Science Foundation. As for the New Zealanders' Scott Base a mile away, Sir Edmund Hillary and his party helped to establish it while on Ross Island in connection with Dr. Vivian Fuchs's Commonwealth Trans-Antarctic Expedition (1957–1958).

21. **"perhaps half a mile"** John Brizindine in *Traveling Times,* November 1977, quoted in P. Mahon, VOE, p. 77.

22. **"the flight deck crew of TE901 took the boss flying"** *Air New Zealand News,* November 30, 1978, quoted in P. Mahon, VOE, p. 113.

NOTES TO CHAPTER 2:
CAPTAIN JIM COLLINS'S FLIGHT

1. **also known as TE901** The prefix "TE," which was Air New Zealand's international flight code until 1990 (when it was changed to "NZ"), is derived from the airline's origin as Tasman Empire Airways Limited (TEAL). Highlights of Captain Collins's flight are according to Vaughan Yarwood, "Erebus: The Fate of Flight 901," *New Zealand Geographic,* November–December 2009, https://www.nzgeo.com/stories/erebus/.

2. **sold out within three weeks** Air New Zealand, *November 1979 Antarctic Flights,* brochure released for November 1979's four flights, New Zealand History, updated November 26, 2019, https://nzhistory.govt.nz/media/interactive/antarctica-flights-brochure.

3. **"Just fabulous," "never seen anything like it," "a cocktail party that ran all day"** *Lookout,* "Flight 901 – The Erebus Disaster," written by Keith Aberdein, and produced and directed by John Keir (New Zealand: TVNZ, 1981), TV series; uploaded to YouTube by mayorofthenonsense as "TVNZ's 'Lookout': The Mt Erebus Disaster," June 15, 2013, https://www.youtube.com/watch?v=yP36X0BsMQ0. For quotations, see 0:01:22, 0:01:26, and 0:02:38 in "TVNZ's 'Lookout.'"

4. **Unlike the merrymaking tourists, TE901's chief purser** Vette and Macdonald, IE, p. 93.

5. **"the carriage of survival suits is unwarranted," "extremely unlikely"** Air New Zealand, quoted in Yarwood, "Erebus: The Fate of Flight 901."

6. **islands whose climate was uncharacteristically mild and humid** Palin, *Erebus: One Ship*, p. 91.

7. **"rolling plateau of what looks like excellent grazing turf"** Vette and Macdonald, IE, p. 123.

8. **"the vast glacier tongues, formed over millenniums"** Vette and Macdonald, IE, p. 126. A section of the Transantarctic Mountains, the Prince Albert Mountains are bounded on the south by the Ferrar Glacier.

9. **"most staunch and willing supporter"** From Edmund Hillary's *No Latitude for Error*, quoted in Carroll du Chateau, "Two Families Share Long Connection," *New Zealand Herald*, January 22, 2008, https://www.nzherald.co.nz/nz/two-families-share-long-connection/W7XYQ4ZK37OJ3JJ5TAHARVBZBY/. The connection would persist even after the deaths in airplane crashes of Hillary's wife (1975) and Peter Mulgrew (1979), for Hillary and Mulgrew's widow would eventually marry (1989).

10. **three Antarctic-themed films** "Antarctic Experience Slide Show," New Zealand History, updated October 4, 2021, https://nzhistory.govt.nz/media/interactive/antarctic-experience. The other two films were *The Big Ice* and *140 Days under the Ice.*

11. **sly episode that occurred during** "Hillary Leads New Zealand Party to South Pole—January 4, 1958," New Zealand

History, updated September 23, 2020, https://nzhistory.govt.nz/page/hillary-reaches-south-pole.

12. **"pretty extensive low overcast"** Communications between local air traffic control and the pilots as well as among those on TE901's flight deck on approach to McMurdo are according to what is recorded on the cockpit voice recorder (CVR), as transcribed by a special New Zealand team working together for a week at National Transportation Safety Board (NTSB) headquarters in Washington, DC. This painstaking transcription was approved by not only the NTSB but also, subsequently, the Privy Council in London. An alternative transcription was later prepared by the New Zealand government's Chief Inspector Ron Chippindale, who, in defiance of international norms, personally made some fifty-odd interpolations or alterations to the text that no one at the NTSB had heard on the tape. Both versions of the transcript can be found in Erebus ("The CVR Transcript"), https://www.erebus.co.nz/The-Accident/Transcript. For Captain Vette's astute commentary on the original CVR transcript's dialogue, see Vette and Macdonald, IE, pp. 142–149.

13. **requested permission from Mac Center to descend** Captain Cooper believes Captain Collins's decision to descend visually was influenced, if not prompted by, the difficulty the pilots were experiencing establishing VHF communication with the Ice Tower; Cooper, "Appendix B," in Vette and Macdonald, IE, p. 255.

14. **"flying with belt and braces caution"** Vette and Macdonald, IE, p. 56.

15. **"you are not looking at what you think you are"** Macfarlane, EP, p. 280.

16. **"I did not have to refer to any of my topographical charts"** Vette and Macdonald, IE, p. 221.

17. **"conflicting inputs"** Vette, "Appendix A," in Vette and Macdonald, IE, p. 240.

18. **US Military Air Command Lockheed C-141 Starlifter** Major Gumble's flight is briefly described in Yarwood, "Erebus: The Fate of Flight 901."

19. **"There was getting to be some suspicion"** Major Gumble, quoted in Erebus ("The Search for TE901"), https://www.erebus.co.nz/The-Accident/The-Accident.

NOTES TO CHAPTER 3:
VANISHED AIRCRAFT

1. **out of contact for over an hour** The accident happened at 1:49:50 p.m. New Zealand summer time, 12:49:50 p.m. McMurdo time, and 00:49:50 a.m. Greenwich Mean Time (GMT).

2. **247 scattered digits** Per Macfarlane, EP, p. 37. Since the Antarctic charters flew a nonstandard route, their pilots were given not a cassette to insert into the onboard computer but a printout whose information had to be laboriously entered into it by hand.

3. **two aviation accident investigators** They were first officer Peter Rhodes for the New Zealand Air Line Pilots Association (NZALPA) and Milton Wylie, the Auckland-based deputy of Ron Chippindale, the head of the government's Office of Air Accidents Investigation.

4. **Then navigation systems specialist Keith Amies phoned** My account of what followed is according to Alan Dorday's statement written two days later (it was *not* made available to the Royal Commission), quoted in Macfarlane, EP, pp. 286–291.

5. **information officer and photographer Nigel Roberts** This account of developments on November 28, 1979, at the two McMurdo research facilities is based on Professor Roberts's

absorbing recollections as delivered in an illustrated lecture at the time of the fortieth anniversary commemorations; see Nigel Roberts, "Remembering the Mt Erebus Disaster," December 3, 2019, video, Victoria University of Wellington, https://vstream.au.panopto.com/Panopto/Pages/Embed. aspx?id=af8516ca-7317-4de5-b31b-ab19001cfd31. I am indebted to Dr. Roberts for bringing this engrossing lecture to my attention (email message to author of February 7, 2022).

6. **"I continued listening at home"** P. Mahon, VOE, p. 14.

7. **Radio New Zealand's political editor was** Paul Holmes, who interviewed Richard Griffin, relates the story in *Daughters of Erebus* (Auckland: Hodder Moa, 2011; hereafter cited as DOE), p. 135.

8. **Deficiencies in the communication apparatus at** Ken Hickson, *Flight 901 to Erebus* (Christchurch: Whitcoulls Publishers, 1980), p. 81.

9. **"That's a question that makes me sick," "We are terrified"** Morrie Davis, quoted in Hickson, *Flight 901*, p. 82.

10. **spotted a suspicious black smear** Erebus ("The Search for TE901").

11. **"Debris at crash site being blown by wind"** Helicopter crew, quoted in Erebus ("The Search for TE901").

12. **"I wanted to tell you that wreckage has been sighted"** Morrie Davis, quoted in Yarwood, "Erebus: The Fate of Flight 901."

13. **"We felt so bad," "playing golf in the shadows"** David Graham, in *White Silence*, ep. 7, "Playing Through."

14. **"The dark cloud of tragedy will be with us all"** Morrie Davis, in *White Silence*, ep. 7, "Playing Through."

NOTES TO CHAPTER 4:
VANISHED SOULS

1. **"I'm at the airport now," "I was really shocked at that"** John Keir and Gary McAlpine, *Erebus Flight 901: Litany of Lies?*, November 2019, produced by New Zealand Media and Entertainment, podcast, MP3 audio, archived by the National Library of New Zealand, https://natlib-primo.hosted. exlibrisgroup.com/primo-explore/fulldisplay?docid=NLNZ_ ALMA11350355710002836&context=L&vid=NLNZ&search_ scope=NLNZ&tab=catalogue&lang=en_US; hereafter cited as *Erebus Flight 901*. See *Erebus Flight 901*, ep. 1, "Champagne Flights to the Ice."

2. **"We couldn't believe what we were hearing"** Elizabeth Collins in a short private memoir, quoted in Holmes, DOE, p. 73. Holmes conducted extensive interviews with Maria Collins and her daughters and spoke with others caught up in the Erebus story in preparation for publishing a book around the time of the disaster's thirtieth anniversary.

3. **"If the plane had landed on the icy sea"** Kathryn Collins, quoted in Holmes, DOE, p. 74.

4. **become "a haunted human being forever"** Maria Collins, quoted in Holmes, DOE, p. 78.

5. **"I always expected that if the worst happened," "I don't even know who it was that rang"** Maria Collins, quoted

in Max Lambert, "Dark Day That Everyone Remembers," *Southland Times*, Stuff, December 3, 2009, https://www.stuff. co.nz/southland-times/news/features/3123531/Dark-day-that-everyone-remembers.

6. **in an odd incident occurring two days after** I thank Stuart Macfarlane for his reconstruction of this otherwise perplexing episode (multiple email message exchanges with author, late May–early June 2022).

7. **They haven't told you about the errors made, have they?** Barney Wyatt, quoted in Vette and Macdonald, IE, pp. 155–156. See also Holmes, DOE, p. 84: "He told Maria the company would 'hang' for the accident because of a 'massive computer error.'"

8. **"was in the Nav Section the following morning," "certainly told Ian Gemmell,' "pretty concerned about what had happened," "pretty sure Keith Amies and the navs," "probably all the navs were there"** David Greenwood in a transcript of a conversation he had thirty years after the accident, in November 2009, with Paul Davison, the lawyer who had represented Maria Collins and NZALPA at the Royal Commission hearings. Since Maria was sent a copy of the transcript, it might appear that she shared it with Paul Holmes in connection with his research into the disaster. In fact, Holmes received the material from Stuart Macfarlane (email message to author, May 30, 2022). Greenwood's remarks are quoted in Holmes, DOE, pp. 93–94.

9. **adversity overcome through sheer willpower** This brief description of Captain Ian Gemmell's backstory is based on

the text of the eulogy delivered by Captain John Gemmell and kindly shared with me (email message to author, April 28, 2022).

10. **"to make sure that he knew [about the coordinate change]"** Greenwood, quoted in Holmes, DOE, pp. 94–95.

11. **corpses "strewn about like they were straws"** Joe Madrid, in *Erebus Flight 901*, ep. 3, "Crash Positions." In this episode, Madrid also says, "I didn't know how bad it had affected me"; Andrew Laxon, "Erebus Flight 901: Litany of Lies. Episode 3: 'I Didn't Know How Bad It Had Affected Me,'" *New Zealand Herald*, November 18, 2019, https://www.nzherald.co.nz/nz/erebus-flight-901-litany-of-lies-episode-3-i-didnt-know-how-bad-it-had-affected-me/C5NJ3WEZ2QO52AKZ7GZ2MZGZWE/.

12. **"Broken bits of everything, including broken bits of bodies"** Hugh Logan, quoted in Holmes, DOE, p. 99.

13. **"tables set up in the gym with food," "this was for the crash survivors," "we knew [they] would not be arriving"** Joseph Madrid, the helicopter's crew chief, quoted in "50th Anniversary of South Pole Flight," New Zealand History, updated September 28, 2021, https://nzhistory.govt.nz/media/photo/50th-anniversary-south-pole-flight.

NOTES TO CHAPTER 5:
RETRIEVED BODIES

1. **"You know," the senior constable remarked** For this quotation, Leighton's volunteering for the Disaster Victim Identification (DVI) squad, and his experience on the ice generally, see Erebus ("Personal Accounts from the Ice"), https://www.erebus.co.nz/The-Accident/The-Recovery-Operation. On Operation Overdue generally, see *Ten One Magazine*, November 28, 2019: "Erebus Part 1: Into the Unknown," https://www.police.govt.nz/news/ten-one-magazine/erebus-part-1-unknown; "Erebus Part 2: The Best of Our Endeavours," https://www.police.govt.nz/news/ten-one-magazine/erebus-part-2-best-our-endeavours; "Erebus Part 3: A Very Important Document," https://www.police.govt.nz/news/ten-one-magazine/erebus-part-3-very-important-document.

2. **"a perfect imprint of the underbelly and wings"** Greg Gilpin in his statement in Erebus ("Personal Accounts").

3. **It has been reliably calculated** On the shock waves, see Vette and Macdonald, IE, p. 153.

4. **"severe storm blew up"** Gilpin, in Erebus ("Personal Accounts").

5. **"It was no game"** Leighton, in Erebus ("Personal Accounts").

6. **describes the grid's layout this way** Sarah Myles, *Towards the Mountain: A Story of Grief and Hope Forty Years on from Erebus* (New Zealand: Allen & Unwin, 2019), Kindle, chap. 8.

7. **"We were assigned a grid," "It was our job"** Leighton, in Erebus ("Personal Accounts").

8. **"penetrates through layers and layers," "hurts your lungs"** Leighton, quoted in Joanna Mathers, "The Secret Scars of Erebus," *New Zealand Herald*, June 28, 2014, https://www.nzherald.co.nz/nz/the-secret-scars-of-erebus/ BAQV4ZGSFCE4XL3XYMZ5OO3724/.

9. **paper identification tag was tied** The all-important ID tags are explained in Myles, *Towards the Mountain*, chap. 8.

10. **"Gee, it's great to be alive"** Leighton, in Erebus ("Personal Accounts").

11. **"We couldn't take them off"** Leighton, quoted in Lena Corner, "AirAsia Flight QZ8501: The Pioneering Work of the Erebus Crash Investigators in 1979 Is Still Being Used Today," *Independent*, January 6, 2015, https://www.independent. co.uk/news/world/airasia-flight-qz8501-the-pioneering-work- of-the-erebus-crash-investigators-in-1979-is-still-being-used- today-9961467.html.

12. **"a gruesome human jigsaw puzzle"** Coroner Allan Copeland, quoted in Hickson, *Flight 901 to Erebus,* p. 190.

NOTES TO CHAPTER 6:
VANISHING EVIDENCE (ANTARCTICA)

1. **having broken away from the rest of the fuselage** Testimony in a statement prepared by Captain Tony Foley of Air New Zealand for Captain Gordon Vette; see Vette and Macdonald, IE, pp. 159–163. Foley was at the crash site in his capacity as an accredited International Federation of Air Line Pilots' Associations air accident investigator.

2. **"virtually unscathed"** Gilpin, in *Erebus: Into the Unknown* (original title *Erebus: Operation Overdue*), directed by Charlotte Purdy and Peter Burger, and written by Peter Burger and Gavin Strawhan (New Zealand: Rogue Productions, 2014), TV movie.

3. **a small, black ring binder with the captain's name** Leighton, in Erebus ("Personal Accounts"); see also P. Mahon, VOE, pp. 268–269. How a petite ring binder whose contents disappeared within two weeks of their discovery came to symbolize the nature of the government's internal inquiry into the accident is described in *Secret New Zealand*, "Erebus Unsolved," written and produced by Jill Graham (New Zealand: Greenstone Pictures, 2003), TV series; uploaded to YouTube by John Parker as "Erebus 901—The Secret Files," August 27, 2019, https://www.youtube.com/watch?v=DpOeL9qkKmY.

4. **"very concerned that the items had not been presented in evidence"** Stanton as described by his close friend John Maine,

who in November 2009 swore an affidavit concerning what Stanton had told him; quoted in Holmes, DOE, p. 114.

5. **NZALPA accident investigator Rhodes had developed grave misgivings** P. Mahon, VOE, p. 156.

6. **seen Captain Gemmell with a bag stuffed with documents** After Rhodes had testified to this effect, Captain Eden, director of flight operations, exerted such pressure on him that he recanted his evidence. In view of not possessing any *tangible* proof, he now declared that he had "no reason to doubt Captain Gemmell in any way, shape or form"; see P. Mahon, VOE, p. 262.

NOTES TO CHAPTER 7:
VANISHING EVIDENCE (NEW ZEALAND)

1. **When interviewed about the Erebus disaster** Ian Gemmell was in his mideighties when film producer Purdy met with him. He died not long after the interview, during which he insisted repeatedly that "the pilots were flying too low"; see "The Making of the Docudrama," "Precis of Research Notes from Meeting Ian Gemmell, May 20, 2012," Erebus Operation Overdue, https://operationoverdue.co.nz/the-making-of.

2. **Muldoon liked his job** Prime Minister Muldoon won three-year terms in 1975, 1978, and 1981. He was voted out of office at a specially called election in July 1984.

3. **"only right and proper," "continuing operational integrity"** Morrie Davis, letter to airline staff in *Air New Zealand News,* quoted in Hickson, *Flight 901 to Erebus,* pp. 199, 200.

4. **grown up flying with her father** I am indebted to the Rev. Dr. Richard Waugh (email message to author, June 22, 2022) for bringing John Munro's aviation career to my attention and directing me to p. 100 of Richard J. Waugh, Bruce Gavin, Peter Layne, and Graeme McConnel, *Taking Off: Pioneering Small Airlines of New Zealand 1945–1970* (Invercargill: Kynaston Charitable Trust, 2003).

5. **"resentful at the nature of the inquiries," "intent on discovering," "everything she told him was"** P. Mahon, VOE, p. 220.

6. **Anne Cassin testified that it was not she** See Anne Cassin in transcript T. 1741, quoted in Macfarlane, EP, pp. 545–546.

7. **In his own testimony before the Royal Commission** Captain Crosbie in transcripts T. 1764 and T. 1765, quoted in Macfarlane, EP, pp. 546–547.

8. **Someone seemed to be going to a lot of trouble** Anne Cassin never spoke publicly about the break-in at her house, possibly since nothing was taken, but Maria Collins was quite voluble about the incident at her residence; see, for example, her recollections in *White Silence,* ep. 1, "The Break-In."

9. **"I required one single file"** CEO Morrie Davis at the Royal Commission hearings, quoted in Macfarlane, EP, p. 542.

10. **Captain Ross Johnson's Antarctic briefing file** P. Mahon, VOE, p. 125.

11. **"the most practical" means of getting rid of records** George Oldfield at the Royal Commission hearings, quoted in Macfarlane, EP, p. 539.

12. **worked at Air New Zealand during the period 1975–1987** "Maurice Williamson MP: 1964 to 1969," Matamata College, https://www.matamatacollege.school.nz/our-alumni/maurice-williamson-mp/.

13. **"Why were people bringing in boxes"** Williamson, quoted in Graham, "Erebus Unsolved."

14. **someone's tie got caught in the shredding machine** Maria Collins in an interview on July 13, 2001, in her Auckland home with Robert Elliott Allinson; cited in Allinson, *Saving Human Lives: Lessons in Management Ethics* (Netherlands: Springer, 2005; hereafter cited as SHL), p. 296, n. 29.

15. **"the Flight Operations Antarctic file, the Navigation Section files"** P. Mahon, VOE, p. 272.

16. **would become suspicious** Mahon in conversation with Macfarlane, cited in EP, p. 538.

17. **"This was at the time the fourth worst disaster"** *Report of the Royal Commission to Inquire into the Crash on Mount Erebus, Antarctica, of a DC10 Aircraft operated by Air New Zealand Limited* (P.D. Hasselberg, Government Printer, 1981; digital copy, https://www.erebus.co.nz/Portals/4/ Documents/Reports/Mahon/Mahon%20Report_web. pdf?ver=2019-06-17-124911-650; hereafter cited as Mahon Report), para. 54. See also Erebus ("The Mahon Report"), https://www.erebus.co.nz/Investigation/Mahon-Report.

18. **"continuing preoccupations," "the veiling of [the airline's] commercial operations"** P. Mahon, VOE, p. 24.

19. **"navigation information and flight plan for the aircraft which crashed"** Morrie Davis, quoted in P. Mahon, VOE, p. 25.

NOTES TO CHAPTER 8:
WHEN ENDS JUSTIFY MEANS

1. **It was "singularly unfortunate"** Peter Mahon in a 1982 speech prepared for aviation lawyers in Sydney, Australia, quoted in Macfarlane, EP, p. 35.

2. **"[I]t has been hard to establish a definite cause"** Arnold Pickmere, "Mt Erebus Investigator Recognized Worldwide," *New Zealand Herald*, February 15, 2008, https://www.nzherald.co.nz/nz/mt-erebus-investigator-recognised-worldwide/OVGR6HF2BAOSMQ4RTYBYUCSAAI/.

3. **the delay was to allow the government's own** P. Mahon, VOE, p. 31.

4. **Unlike in England** See Macfarlane, "Notes on Text," in Vette and Macdonald, IE, p. 325.

5. **"counterpunching" active and potential opponents** For Muldoon's abrasive personality, autocratic and combative manner, aggressive style, and fondness for counterpunching, see Simon Walker, "Obituary: Sir Robert Muldoon," *Independent*, August 5, 1992, https://www.independent.co.uk/news/people/obituary-sir-robert-muldoon-1538341.html.

6. **aviation writer working in-house for Air New Zealand** "The Hickson Team Is Back," Avenue for Creative Arts, October 18, 2017, www.fifthavenue.asia/326-2/.

7. **"didn't think an [additional] inquiry," "methods and expertise,"
"had complete faith"** Hickson, *Flight 901 to Erebus*, p. 221.

8. **chief inspector's "competence and authority,"** Hickson, *Flight 901 to Erebus*, p. 232.

9. **Chippindale would share their contents** From Ian Gemmell's testimony at the Royal Commission hearings, quoted in Holmes, DOE, p. 311.

10. **"Jim was too low, Maria"** Chippindale's remark as recalled by Mrs. Collins and quoted in Holmes, DOE, p. 158.

11. **"the last thing which occurred in the sequence"** Chippindale at the Royal Commission hearings, quoted in Holmes, DOE, p. 150.

12. **their computerized Antarctic track would be over Mt. Erebus** *Aircraft Accident: Air New Zealand McDonnel-Douglas DC10-30 ZK-NZP, Ross Island, Antarctica, 28 November 1979, Report 79-139* (Office of Air Accidents Investigation, Ministry of Transport, 1980; digital copy; hereafter cited as Chippindale Report), https://www.erebus.co.nz/Portals/4/Documents/Reports/Chippindale/79-139%20Chippindale%20Report%20-%20Web.pdf, para. 1.1.1. See also Erebus ("Chippindale Report"), https://www.erebus.co.nz/Investigation/Chippindale-Report.

13. **"He told me the route he was going to take," "It wasn't where he said he was going," "the Dry Valleys, the coast of Victoria Land"** Kathryn Carter (née Collins) in an interview with Holmes, DOE, p. 361.

14. **"fairly close to this bumpy lot"** Elizabeth Collins in her official affidavit for the Royal Commission, quoted in P. Mahon, VOE, p. 216.

15. **did not constitute "evidence" as he understood that term** At the Royal Commission hearings, Chief Inspector Chippindale did not contest the veracity of Mrs. Collins's testimony or that of her daughters' affidavits. In a venomous press release *after* the Mahon Report was published, however, he would accuse the three women of fabricating the story that Captain Collins was working on maps the evening prior to the fatal flight; Chippindale press release of February 3, 1982, quoted in Macfarlane, "Notes on Text," in Vette and Macdonald, IE, pp. 327–328.

16. **alleging Captain Collins had no legitimate checking reference** Chippindale maintained that Captain Collins could not have acquired a large topographical map of Antarctica because "they were not readily available over the counter" in New Zealand (Chippindale press release of February 3, 1982, quoted in Macfarlane, "Notes on Text," in Vette and Macdonald, IE, p. 327). However, as an international pilot, Captain Collins would obviously have had ample opportunity to pick up this type of specialized product in his travels. In fact, his spare copilot, Graham Lucas, had borrowed a copy of GNC21N from a friend and colleague, helicopter pilot Peter Tait, as the chief inspector well knew since he had been the first to discover that (see Holmes, DOE, pp. 368–369). As for the small map, NZMSB5, Chippindale had early on discovered its identity too. Although it was readily available locally, the chief inspector publicly described it as extremely difficult to come by.

17. **impossible for Captain Collins *not* to know where he was**
The horizontal situation indicator (HSI) is briefly explained in
Vette and Macdonald, IE, pp. 20, 114–115.

18. **suddenly afflicted by a mystery illness** For Justice Mahon's
questioning of Captain Kippenberger with respect to the
possible psychological or medical impairment of the pilots,
see Macfarlane, EP, p. 399.

19. **believed the infallible navigation system** Prior to the crash
on Erebus, it was considered unimaginable that pilots could *ever*
be shown a flight plan at briefing and then be given a different
one for insertion into their aircraft's computer without being
notified. Effectively, the undisclosed substitution rendered
Captain Collins's just-plotted track tragically outdated.

20. **recovered and delivered to Auckland** See "Recovering the
Flight Recorders from Erebus," New Zealand History, updated
September 19, 2019, https://nzhistory.govt.nz/media/photo/
recovering-flight-recorders-erebus.

21. **quality of the recording proved abysmal** Captain Cooper's
description of the laborious transcription process at the NTSB
appears as "Appendix B" in Vette and Macdonald, IE, pp. 244–257.

22. **Chippindale now took sole custody** On the liberties the
chief inspector took with the audiotape—inviting Gemmell to
listen to it alone with him, taking it to the United Kingdom to
reanalyze exclusively in the company of Farnborough officials,
and personally editing it—see Gary Parata's fine article, "The
CVR Transcript Controversy," Erebus, https://www.erebus.
co.nz/Investigation/The-CVR-Transcript-Controversy.

23. *fifty-five deviations* from the Washington Transcript Captain Cooper summarized the effect of the many subtle changes as being "detrimental to the crew"; Cooper, quoted in "Flight Transcript: The Moment of Impact," National, Stuff, November 20, 2009, https://www.stuff.co.nz/national/erebus/3030756/Flight-transcript-The-moment-of-impact).

24. **importuned them with "mounting alarm"** Chippindale Report, para. 2.20; see also paras. 2.25 and 3.24.

25. **"the purpose of the transcript," "support the conclusions of the report," "continued editing the transcript"** From Chippindale's testimony at the Royal Commission hearings, quoted in Macfarlane, "Notes on Text," in Vette and Macdonald, IE, p. 343.

26. **"Taylor on the right now"** From "Original Washington Transcript," Erebus ("The CVR Transcript").

27. **"The Taylor or the Wright now or do yah?" "No, I prefer here first!"** From the Farnborough transcript of the CVR. Taylor and Wright were two of the three Dry Valleys. For Chippindale's CVR transcript, see https://www.erebus.co.nz/The-Accident/Transcript.

28. **"unable to identify any place with the single name of Wilson"** Chippindale's testimony at the Royal Commission hearings, quoted in Macfarlane, "Notes on Text," in Vette and Macdonald, IE, p. 344.

29. **"Bit thick here, eh Bert"** While conceding there was no one on the flight—actually, on the entire aircraft—named Bert,

Chippindale insisted "that is far from proof that no one called 'Bert' was on the flight deck, neither is it in any way proof that the words were not spoken"; Chippindale in a press release quoted in Macfarlane, EP, p. 329.

30. **Perhaps "Bird," as in Cape Bird** Although "This is Cape Bird" did not meet the stringent criteria for inclusion in the Washington Transcript, that assertion at that exact point in the aircraft's journey would not have been out of place. In fact, because he knew the precise time of impact, the precise time the remark was made, and the precise speed of the aircraft, Justice Mahon was able "to fix the geographical point at which Mulgrew had spoken" (VOE, p. 196). He consequently agreed with the view of Colonel Turner of the NTSB that the commentator saw Cape Tennyson appear and took it to be Cape Bird (VOE, p. 197).

31. **particularly misleading ancillary charge** For a discussion of the efficacy of radar in the presence of snow and ice, see P. Mahon, VOE, pp. 165–169.

32. **"inability of the radar pulses from the radar to achieve"** P. Mahon, VOE, pp. 168–169.

33. **Before Mahon left Bendix's premises** P. Mahon, VOE, pp. 169–170.

34. **"it was conducted in public"** Macfarlane, "Notes on Text," in Vette and Macdonald, IE, p. 301.

35. **chatting "informally and off the record," "the maximum amount of information"** Chippindale's testimony at the Royal

Commission hearings, quoted in Macfarlane, "Notes on Text," in Vette and Macdonald, IE, p. 303.

36. **"some eighteen highly qualified international experts"** Chippindale in a press release dated January 18, 1982, quoted in Macfarlane, "Notes on Text," in Vette and Macdonald, IE, p. 301.

37. **stressed that he had not been under pressure** Chippindale in a press release dated February 3, 1982, cited in Macfarlane, "Notes on Text," in Vette and Macdonald, IE, p. 306.

38. **"based on the assessment made by all the investigators"** Chippindale in *The Auckland Star*, January 29, 1982, quoted in Macfarlane, "Notes on Text," in Vette and Macdonald, IE, p. 306.

NOTES TO CHAPTER 9:
QUESTING FOR TRUTH

1. **"I was stunned," "I had flown with every crew member"**
Vette and Macdonald, IE, p 25.

2. **decided then and there** Vette felt a "very definite compulsion" to undertake his own investigation once he learned where the government's inquiry was headed. "I don't really think that I go around looking for fights or arguments, particularly when it obviously meant coming out of my sinecure at the top of the seniority list." However, "the philosophy being espoused by the company and the department [Civil Aviation]" repulsed him to the point where he "determined to act"; see Gordon Vette and John Macdonald, *Impact Erebus Two*, a documentary that includes interviews of Vette and Mahon together in 1984 (2000; uploaded to YouTube by mayorofthenonsense, June 15, 2013, https://www.youtube.com/watch?v=xyWvOI_MD-Q). Later, when the government's chief inspector of air accidents completed his interim report on the disaster, Vette would find himself in "complete disagreement" with its conclusions; Vette and Macdonald, IE, p. 27.

3. **"carried out between flying assignments"** Vette and Macdonald, IE, p. 26.

4. **"extraordinary degree of incompetence"** Vette and Macdonald, IE, p. 35. When he learned that "Captain Collins had flown for at least twenty-five miles in Nav Mode and in

VMC [visual] conditions straight into Mt. Erebus," Vette ruled out "the possibility of drunkenness, drugs, uncharacteristic behavior, suicidal tendencies" because of "the fail-safe crew concept"; Vette, "Appendix A," in Vette and Macdonald, IE, pp. 236–237.

5. **accepted as "axiomatic," "even an experienced air crew"** Vette and Macdonald, IE, p. 26.

6. **commendation from McDonnell Douglas for "the highest standards"** Wikipedia, s.v. "Cessna 188 Pacific Rescue," last modified September 2, 2021, https://en.wikipedia.org/wiki/Cessna_188_Pacific_rescue.

7. **recounted in a book and dramatized in an American TV movie** The 1989 book is Stanley Stewart's *Emergency! Crisis on the Flight Deck*. The 1993 movie is *Mercy Mission: The Rescue of Flight 771*.

8. **"innocent victims, trapped," "public had been soaked," "old friends and colleagues"** Vette and Macdonald, IE, pp. 27, 28, 29.

9. **"CRASH REPORT POINTS TO ERROR," "THOUSANDS OF FEET TOO LOW," "CREW UNCERTAIN"** Quoted in P. Mahon, VOE, p. 41.

10. **"had not misled the aircrew," "poor surface and horizon definition"** Chippindale Report, paras. 3.7, 3.37.

11. **the "finest passage" of *The Divine Comedy*** Peter Mahon in a 1977 letter to his daughter, Janet, quoted in S. Mahon, MFS, pt. 1, "Exhibit C: *Dante the Maker*." In *Dante* (New York: Viking

Penguin, 2001; p. 200), R. W. B. Lewis calls *Dante the Maker*, by William Anderson, "the best and most thorough biography in English." A devoted afficionado of the Italian poet, Justice Mahon was once invited to critique Anderson's complex work, although he never did manage to do so.

12. **abundant testimony by line pilots to the contrary** Captain White, for instance, had been standby captain in 1978 as well as captain in command of the 1979 flight the week before Captain Collins's fatal tour. As such, he had attended two briefings, at both of which Captain John Wilson was the briefing officer and at the latter of which Captain Ross Johnson also was present. "No mention was made by either Captain Wilson or Captain [Ross] Johnson to the effect that Mt. Erebus lay directly on track" (White's testimony at the Royal Commission hearings, quoted in Macfarlane, EP, p. 270). He understood the route to run up McMurdo Sound—to the *west* of Mt. Erebus, not *over* it.

13. **"not merely the evidence given but the way it was given"** Peter Mahon in a letter to James Buckley QC, quoted in S. Mahon, MFS, pt. 3, "After Erebus."

14. **nonverbal signs are quite conspicuous** Brett Christmas, the son of one of TE901's victims: "[A]s a physician I know how important non-verbal cues are in interpreting the credibility of an account"; email message to author, February 1, 2022.

15. **took issue with the briefing officers' claims** See P. Mahon, VOE, pp. 121–124.

16. **"That never happened"** John Gabriel to Arthur Cooper, quoted in Holmes, DOE, p. 255.

17. **"extreme fine detail [Captain Wilson] appears to have included"** Les Simpson's testimony at the Royal Commission, quoted in Macfarlane, EP, p. 687. On the impressiveness of Captain Simpson as a witness, see P. Mahon, VOE, pp. 121–122.

18. **"most consummate confusion," "make the South Magnetic Pole"** P. Mahon, VOE, pp. 130, 87.

19. **Captain Gemmell quietly obtruded himself into the mix** For Captain Keesing's original negotiations on behalf of the airline with Civil Aviation regarding the altitude and flight path of the envisioned Antarctic aerial tours, see P. Mahon, VOE, pp. 127–130. Captain Keesing did not learn that his plans had been discarded until the Royal Commission hearings. Ironically, his own low-flying aerial tour at 2,000 feet up McMurdo Sound had served to confirm to him that his original plans were being honored.

20. **"I fail to understand why," "follow the operational route"** Brief of Robert Baden (Bob) Thomson, superintendent of the Antarctic Division of the Department of Scientific and Industrial Research (DSIR) and four-time flight commentator (including the inaugural flight). Bob Thompson had visited Antarctica over fifty times and had spent five years there; Supplementary Statement by R. B. Thomson, in Stuart Macfarlane, Unpublished Papers.

21. **would describe in an affidavit** Mahon Report, para. 148.

22. *no sightseeing DC-10s ever overflew* Captain Vette confirmed this in a private interview on May 13, 2001, with author Robert Elliott Allinson, who references it in *Saving Human Lives*, p. 291, n. 23. He also confirmed it with Stuart Macfarlane, who was on

excellent terms with Captain Vette and even ghostwrote two late additions to *Impact Erebus* at Vette's request (email message to author, June 1, 2022). Macfarlane became involved when the book was already in galley proof form and was responsible for the pictorial section and the epilogue.

23. **shifted his route horizontally** Peter Grundy, Brief of Evidence: "I did not track direct to Williams Field but down the Victoria Land coast to the west of the direct path"; in Macfarlane, Unpublished Papers.

24. **Captain Vette likewise abandoned the nav track** Vette, "Appendix A," in Vette and Macdonald, IE, p. 220.

25. **"deviating slightly whilst approaching Ross Island"** Ian Gemmell, Brief of Evidence, in Macfarlane, Unpublished Papers. In a rare mistake probably owing to his writing fast and from memory, Justice Mahon (VOE, p. 83) describes Captain Gemmell as testifying that he had flown over the summit of Mt. Erebus just east of its peak (incorrect) instead of deviating horizontally from it (correct and in his Royal Commission testimony). That Captain Gemmell maintained the minimum safe altitude prescribed for flights over the volcano (even though he did not overfly it) is confirmed by a passenger who took photographs on Air New Zealand's inaugural aerial tour. Tourist John McCombe of the *Christchurch Star* reported that Captain Gemmell was flying at an altitude 3,000 to 4,000 feet higher than the mountain; Hickson, *Flight 901 to Erebus*, p. 98.

26. **"would not have overflown the plume of the volcano because"** Vette, "Appendix A," in Vette and Macdonald, IE, p. 220. The volcano was then active.

27. **in Cooper's experience, Hewitt made mistakes** For Arthur
Cooper's recollections regarding Hewitt, see Holmes, DOE,
pp. 241–242.

28. **Evidence that Air New Zealand** The evidence was a statement
written by Flight Dispatch's Dorday two days after the accident.
The statement made clear that Hewitt was completely unaware
of his programming mistakes. The airline revealed the existence
of Dorday's statement only later for the purpose of using it,
during the Court of Appeal and Privy Council proceedings, to
substantiate Air New Zealand's claim that there had been no
organized attempt at perjury.

   Dorday's programming errors were (1) to move the airline's
destination waypoint west twenty-seven miles and (2) to move
it from Williams Field (destination waypoint #1) rather than
the nondirectional beacon (NDB; destination waypoint #2).
This indicates a hand—not his own—had made the switch to
the NDB in mid-1977 without informing him.

29. **Exhibit 164 demonstrated** On Exhibit 164 and Annex J, see
Vette and Macdonald, IE, pp. 44–45.

30. **considerable damage was being done** According to Justice
Mahon's son Sam, "My brother, who had been visiting the
hearings regularly, reported that from time to time the whole
court would be trembling with mirth at some of the evidence
given, it was so palpably unbelievable"; S. Mahon, MFS, pt. 1,
"Exhibit E: The Table on which He Wrote *Verdict on Erebus*."

31. **neither could say "a bloody thing"** Morrie Davis spoke for
thirty-five of the forty-minute exchange, urging Cooper and

Rhodes not to fret about what was happening at the Royal Commission hearings because there was "a bigger picture going on"; Cooper, quoted in Holmes, DOE, p. 343.

32. **"the area they flew into [Lewis Bay] bore an uncanny resemblance"** Vette and Macdonald, IE, p. 169.

33. **"only a narrow sector of the total horizon," "atmospheric transparency"** Vette and Macdonald, IE, p. 213.

34. **"the forward vista of snow-covered terrain was rising"** P. Mahon, VOE, p. 153. Chief pilot Wilson of Helicopters New Zealand Limited gave an account of the weather pattern in the vicinity of Ross Island as observed by him in late December 1979 through early January 1980—that is, in the wake of the accident. There was a predictable sequence: cloud-clear mornings, fog rolling in over Mt. Erebus's lower slopes at about 11:00 a.m. and gradually rising upward while simultaneously the middle of the mountain became covered in cloud. By early afternoon the cloud would extend to its summit, "and the fog and cloud base would merge at a point 1,000 feet to 1,500 feet above sea level"; Wilson, "Appendix D," in Vette and Macdonald, IE, p. 265. What Captain Vette termed a "ramp effect"—fog rolling in from Lewis Bay and proceeding up the crash slope to obscure the ice cliffs—was captured on camera by Erebus air crash investigator Captain Foley; quoted in Vette and Macdonald, IE, p. 163.

35. **"flat-light illusion"** This optical deception is not uncommon in the polar regions, northern Canada, and northern Europe. Although in whiteout a dark object will be visible for miles, "a snow-covered object, even a mountain next to the observer,

will be invisible"; Vette, "Appendix A," in Vette and Macdonald, IE, p. 234. The essence of the illusion is that snowy, rising terrain will become invisible insofar as human perception is concerned; the rising ground "flattens out" to human eyes, leaving the observer to believe the terrain flat for many miles into the distance; P. Mahon, VOE, pp. 42, 151.

36. **"gives *no depth perception"*** Rhodes, "Appendix E," in Vette and Macdonald, IE, p. 271 (emphasis added). For Rhodes, the generic term "whiteout" was a deceptively tame word to describe Antarctica's pernicious varieties of visual illusions. He referred to the general phenomenon as the "flying in a milk bottle" effect and said that effect was worst when the sky was overcast.

37. **identified the capes to the left and right as being those at the entry** A written statement that Chippindale had first prepared for the minister of transport after the publication of the Mahon Report (1981) was released in February 1982. In it the chief inspector claimed that there could be no excuse for the crew's descent below 16,000 feet without first establishing the aircraft's position as accurately as possible. Vette's fastidious research confirms that the airmen had, indeed, fixed their position prior to descent. They were wrong for the reasons he cites, but they *believed* they knew exactly where they were.

38. **"they would subtend exactly the same retinal angle"** Vette, in Vette and Macdonald, *Impact Erebus Two.*

39. **"the two thin strips of dark rock," "if the Captain's Nav Track confirmed"** P. Mahon, "Appendix F," in Vette and Macdonald, IE, p. 285.

40. **Beaufort Island is seven square miles** Knowing firsthand how proficient commentator Mulgrew was at identifying, *from the air,* Antarctic topographical features, or landmarks, Captain Vette was led to conclude that Beaufort Island was simply "never available to him for identification" on the day of the fatal flight; Vette, "Appendix A," in Vette and Macdonald, IE, p. 221.

41. **the aircraft was just then banking left** For Justice Mahon's scrupulous reconstruction of and calculations regarding this critical stage in the fatal flight, see VOE, pp. 201–202, and Vette and Macdonald, IE, p. 54.

42. **until fifty-five seconds before impact** Cooper, "Appendix B," in Vette and Macdonald, IE, p. 249.

43. **could still not reach the Ice Tower on VHF radio** Cooper offers four hypotheses to explain why the pilots were merely puzzled, not alarmed, by their failure to connect with the Ice Tower; Cooper, "Appendix B," in Vette and Macdonald, IE, p. 255.

44. **"You have an ordinary overcast"** P. Mahon, in Vette and Macdonald, *Impact Erebus Two.*

45. **Mahon later retraced Captain Collins's exact flight track** P. Mahon, "Appendix F," in Vette and Macdonald, IE, pp. 293–294.

46. **"Suppose that there had been no discussion"** P. Mahon, VOE, p. 199.

47. **"200 times greater than that of gravity"** Vette and Macdonald, IE, p. 81.

NOTES TO CHAPTER 10:
"THE HOVERING FATES"

1. **disengaged from the navigation track at Cape Hallett**
   See Captain Simpson's testimony at the Royal Commission hearings regarding the last stage of his outbound flight two weeks before Captain Collins's flight, quoted in Macfarlane, EP, pp. 242–243.

2. **phoned Captain Ross Johnson, flight manager (line operations)** See Simpson's testimony at the hearings, quoted in Macfarlane, EP, p. 243.

3. **"Captain Simpson rang me and said"** From Captain Johnson's statement of December 10, 1979, quoted in Macfarlane, EP, p. 252.

4. **at the Royal Commission hearings, he revised his recollection** From Captain Johnson's testimony at the Royal Commission hearings, quoted in Macfarlane, EP, 252; see also P. Mahon, VOE, pp. 136–137.

5. **"did not report this matter to Captain [Ross] Johnson as an error"** From Captain Simpson's testimony at the Royal Commission hearings, quoted in Macfarlane, EP, p. 243. Simpson went on to note that "if I had thought there was an error involved I would have included it in my flight report and would have advised the nav section in the same manner as I have reported other matters on previous occasions." In other words, if Captain Simpson was telling the truth, the destination

waypoint could not possibly have been changed this last fatal time because of Captain Simpson's input.

6. **he and his colleague had instead consulted** Nicholson's questioning of Hewitt on this point at the Royal Commission is reproduced in Macfarlane, EP, pp. 259–260.

7. **Did the man have amnesia?** If Hewitt was telling the truth when he both wrote and testified that he had no inkling he'd moved the destination waypoint twenty-seven miles rather than two and one-tenth miles eastward in November 1979, it must follow that upon moving it twenty-seven miles westward over a year earlier, he had never proceeded to double-check that he'd correctly programmed that waypoint's coordinates into the new ground computer either. It was true that from just viewing the screen in front of him he could not discover his 1978 and 1979 errors. He needed to print out a copy, a commonsense procedure, and check any proposed change against it. Despite the possible ramifications a mistake could have on a distant journey to a forbidding and dangerous part of the globe, Hewitt did not print out a copy either year.

8. **"exactly where the blame lay"** See Cooper's remarks quoted in Holmes, DOE, p. 242.

9. **chief traffic controller and supervisor of Mac Center would** P. Mahon, VOE, p. 159.

10. **Whether by accident or design** The Navigation Section officer responsible for the change was surnamed Brown. For Justice Mahon's reasoning for disbelieving him, see the Mahon Report, para. 247.

11. **We know the airline's transponder received a radar signal** Vette in a 2001 interview with Allinson, cited in SHL, p. 307.

12. **the official who ran Mac Center believed the DC-10 had not** P. Mahon, VOE, p. 159.

13. **told the tape's last portion, no difficulty imagining** P. Mahon, VOE, pp. 160–162.

14. **"the U.S. military put up a protective shield"** Macfarlane in an email to author (May 30, 2022) that describes his visit to California to speak with the plaintiffs' attorneys. For the radar operator's alleged mental breakdown after the crash of TE901, see Allinson's 2001 interview with Macfarlane, quoted in Allinson, SHL, p. 307.

15. **Captain White disengaged his nav track** See White's statement reproduced in Macfarlane, EP, p. 273.

16. **regularly unlocked from the automatic navigation system** P. Mahon, "Appendix F," in Vette and Macdonald, IE, p. 275. See also Vette and Macdonald, IE, p. 85.

NOTES TO CHAPTER 11:
THE MAHON REPORT'S
CONCLUSIONS ON CAUSATION

1. **"his one truth"** S. Mahon, MFS, pt. 4, "Art, Religion, and Philosophy."

2. **"In my life as a kid"** S. Mahon, quoted in Jarrod Booker, "Mahon on Mahon," *Otago Daily Times*, September 13, 2008, https://www.odt.co.nz/lifestyle/magazine/mahon-mahon.

3. **"put me and counsel to the Royal Commission onto the right path"** P. Mahon, in Vette and Macdonald, *Impact Erebus Two*.

4. **"small acts, errors, or omissions," "when combined, form a trap," "the flying in cloud, pilot error theory," "later very angry about that"** Vette, in Vette and Macdonald, *Impact Erebus Two*.

5. **"there were a lot more casualties out of Erebus than"** Vette, quoted in "Court Action Following Erebus Disaster Inquiry," New Zealand History, updated September 18, 2019, https://nzhistory.govt.nz/media/photo/ongoing-debate-about-erebus-disaster.

6. **Wilson startlingly revealed** Justice Mahon attributed Captain Wilson's stunning admission to an underlying strength of character. That strength of character, he said, was demonstrated

during World War II and had won him the Air Force Cross (VOE, p. 115).

7. **"cling[ing] like a leech"** P. Mahon, VOE, p. 234.

8. **they expressed incredulity** P. Mahon, VOE, p. 103.

9. **he himself would have recognized** P. Mahon, VOE, p. 104.

10. **They'd *all* been put at risk** Vette, in Vette and Macdonald, *Impact Erebus Two.*

11. **"any culpable act or omission"** Mahon Report, para. 386.

12. **how "unintelligent and obtuse," "brazen the whole thing out to the end"** P. Mahon, VOE, p. 251.

13. **In a provocative mental exercise** For Justice Mahon's musings, see P. Mahon, VOE, p. 278.

14. **"cut through argument and humbug"** McNeish, BR, p. 219.

15. **"He simply turned towards me and spread his arms outwards"** P. Mahon, VOE, 229. On the tenth anniversary of the Erebus disaster, Chief Inspector Chippindale admitted in an interview that the 16,000 feet minimum safe altitude dictum was a smokescreen by which Air New Zealand hoped to avoid insurance liability. He conceded that the airline had, in fact, condoned low flying on approach to McMurdo. "Erebus Crash: Myths and Reality," *Dominion Post*, Stuff, January 31, 2009, https://www.stuff.co.nz/dominion-post/archive/national-news/265485/Erebus-crash-myths-and-reality.

16. **"and was known by those in the company to lie"** P. Mahon, VOE, p. 241.

17. **identified ten contributing causes** Mahon Report, para. 387.

18. **the one that "continued to operate from before the aircraft left"** Mahon Report, para. 392.

19. **These errors have been succinctly described** See Macfarlane, "Notes on Text," in Vette and Macdonald, IE, pp. 329–340.

20. **"I was quite unable to accept all of these mistakes"** P. Mahon, VOE, p. 247. In a letter to James Buckley QC, Justice Mahon put it this way: "All the navigation witnesses were persistently cross-examined before the Royal Commission on the basis that a twenty-seven-mile shift in the destination waypoint was known to the Navigation Section to have occurred in 1978, and that evidence to the contrary was so improbable as to be without credibility." He went on to note that "once I had rejected the thesis of all the cumulative mistakes then I had no alternative but to express the view that there had been a combined attempt by the Navigation Section to deceive the Royal Commission"; quoted in S. Mahon, MFS, pt. 3, "After Erebus."

21. **management system that depended on verbal communications** P. Mahon, VOE, p. 227. See also the Privy Council's comments in Mahon v. Air New Zealand Ltd [1984] 3 All ER 201; [1983] NZLR 662, http://www.nzlii.org/nz/cases/NZPC/1983/3.html, s. 27.

22. **Air New Zealand was riddled with substandard** Despite receiving assurances that any minor issues at the airline had been

resolved, Justice Mahon would learn of yet another computer error a year after the Erebus disaster. This one also involved a faulty flight plan for an overseas trip. While Flight Dispatch caught the problem before takeoff, Justice Mahon was deeply troubled by this latest example "of the lack of checking and the lack of communication which had been the distinguishing feature of the Flight Operations Division as revealed by the evidence at the inquiry"; P. Mahon, VOE, p. 232.

23. **"By a navigational error for which the aircrew was not responsible"** P. Mahon, VOE, p. 295.

NOTES TO CHAPTER 12:
THE ROYAL COMMISSIONER'S
CREDIBILITY FINDING

1. **weighed whether to ignore ordinary judicial protocol** P. Mahon, VOE, pp. 248–250.

2. **"No judicial officer ever wishes to be compelled"** P. Mahon, VOE, p. 249.

3. **Margarita Mahon would explain** Margarita shared her memory of the circumstances surrounding her husband's crafting of "an orchestrated litany of lies" in *White Silence,* ep. 8, "The Apology."

4. **"a man of integrity who was nobody's lackey"** McNeish, BR, p. 224.

5. **"an uncompromising adversary," "never saw him back down"** S. Mahon, MFS, pt. 2, "Album," "Folks on the Hill."

6. **"This might have been one of those occasions"** Ted Thomas's comments in this paragraph are from "Erebus Flight 901: Litany of Lies. Episode 9: The Five Words That Sank Justice Peter Mahon," *New Zealand Herald,* November 26, 2019, https://www.nzherald.co.nz/nz/erebus-flight-901-litany-of-lies-episode-9-the-five-words-that-sank-justice-peter-mahon/DWSLQL4FXWGA7FXFQQ2N2AGOOQ/. On his evaluation of the "five words," see also *Erebus Flight 901,* ep.

9, "Report and Reaction." See also Prime Minister Muldoon's view: "It would have been better if Mr. Justice Mahon had phrased his views less elegantly and more precisely"; Muldoon in *New Zealand Truth*, January 12, 1982, quoted in Barry Gustafson, *His Way: A Biography of Robert Muldoon* (Auckland: Auckland University Press, 2000), p. 291.

7. **"determined that he should pay for it"** Gary Harrison, in *Erebus Flight 901*, ep. 10, "The Years Since."

8. **never properly explained the "litany of lies" because** This intriguing explanation was offered by University of Auckland law professor Bernard Brown to his friend Stuart Macfarlane, who thinks it is probably correct; Macfarlane, email message to author, February 11, 2022.

9. **stealthily arranged for the police** See chap. 13.

10. **"I had to make it quite clear"** Justice Mahon in a Radio New Zealand interview with Sharon Crosbie, as reported in the *New Zealand Herald*, February 2, 1982, and quoted in Macfarlane, EP, p. 521.

11. **two sides of the same coin** Macfarlane, EP, p. 518.

12. **"I don't believe a Commission of Inquiry"** Justice Mahon in an interview in *New Zealand Listener*, March 13, 1982, quoted in Macfarlane, EP, p. 517.

NOTES TO CHAPTER 13: SPARRING AFTER THE MAHON REPORT'S RELEASE

1. **"The findings do not accurately appear to be in accord," "In other words"** Rob Muldoon, in *Erebus Flight 901*, "Report and Reaction."

2. **"Prime Minister, you've attacked this Royal Commissioner," "That is what we thought at the time"** See Justice Mahon's remarks in Vette and Macdonald, *Impact Erebus Two*.

3. **"We should have appointed more than one judge"** Rob Muldoon, quoted in Holmes, DOE, p. 382.

4. **"I don't think two other commissioners would ever," "leavened some of Mahon's language," "It was in my view if not inappropriate then certainly unusual"** McLay, in "White Silence: Former Deputy Prime Minister's Biggest Regret after Air New Zealand Erebus Crash," Stuff, https://www.stuff.co.nz/national/117079204/white-silence-former-deputy-pms-biggest-regret-after-air-nz-erebus-crash.

5. **"looking after a clutch of personal political friends," "thought from day one"** Bill Rowling, in *White Silence*, ep. 4, "Love Song."

6. **"What we are going to do is"** Muldoon, in *White Silence*, ep. 4, "Love Song."

7. **dismissed the Mahon Report's findings as "unsupported"** Muldoon, as quoted in *Erebus: The Aftermath*, directed by Peter Sharp, written by Greg McGee, and produced by Caterina De Nave (New Zealand: TVNZ, 1987), TV miniseries; see "'Erebus—The Aftermath' Part Two," uploaded by John Parker, June 21, 2016, YouTube video, https://www.youtube.com/watch?v=avnBAtLvVqY&t=5s. The docudrama's dialogue hews very closely to what the speakers involved are known to have said to protect Television New Zealand (TVNZ) against defamation charges. Meticulously, playwright Greg McGee had "recorded the origin of every bit of information, every scene and virtually every line of dialogue attributed to any character"; see Greg McGee, *Tall Tales (Some True): Memoirs of an Unlikely Writer* (New Zealand: Penguin Books, 2008; hereafter TT), p. 256. He told actors that if they stuck strictly to the scripted text, they'd be covered against any defamation charges. As additional insurance, the dialogue was vetted by a top defamation lawyer.

8. **Justice Mahon called a press conference** What was said at it is according to *Erebus: The Aftermath*.

9. **"Once, while on patrol in Italy"** Justice Mahon in a letter to Sian Elias, quoted in S. Mahon, MFS, pt. 4, "Liberalism, Reactionism, and the Inarticulate Premise."

10. **walking onto a minefield to retrieve a friend whose feet** Booker, "Mahon on Mahon."

11. **he denounced its "totally indefensible contents," "reject entirely"** Morrie Davis in *White Silence*, ep. 4, "Love Song."

12. **"remove a focus point from the current controversy"** "Court Action Following Erebus Disaster Inquiry," New Zealand History.

13. **a member of the airline's cabin crew** For Pogo Watson's discreet insights, see *White Silence*, ep. 7, "Playing Through."

14. **did not inform Muldoon of his action** Jim McLay, in *Erebus Flight 901*, "Report and Reaction."

15. **The first person the police called in for questioning** The testimony of Brian Hewitt and Captain Ross Johnson is according to *Erebus: The Aftermath*.

16. **the accident was attributable not so much to any individuals** Mahon Report, para. 393.

17. **documentary on the Erebus disaster** See mayorofthenonsense, "TVNZ's 'Lookout,'" 45:30.

18. **"navigation data which Captain Collins was given"** P. Mahon, VOE, p. 121.

19. **"more heartache and stress," "because I wanted to see justice," "not dealt with correctly"** Gilpin, quoted in Holmes, DOE, p. 123. See also Gilpin's declaration in *Operation Overdue*: "I believe in honesty. I wanted to see justice being done."

20. **regarding extramarital lovers** See "Making of the Docudrama," "Précis of Research Notes," Erebus Operation Overdue.

21. **"references of a general nature such as shopping lists"**
Crosbie's statement, quoted in Macfarlane, EP, p. 551.

22. **"honest to the last part of his being," "*he* knew the difference between a shopping list and co-ordinates"** Maria Collins, quoted in Tim Donoghue, "Family Pays Tribute to Erebus Officer," *Dominion Post*, Stuff, November 12, 2011, https://www.stuff.co.nz/dominion-post/news/5953754/Family-pays-tribute-to-Erebus-officer (emphasis added).

23. **an old friend of Chief Superintendent Brian Wilkinson**
Holmes, DOE, p. 123. Justice Mahon had shared with Stuart Macfarlane that Wilkinson had told him that Prime Minister Muldoon was taking steps to kill Wilkinson's investigation in the crib—and Macfarlane later shared that intelligence with Holmes.

24. **"significant shortfalls in the overall performance"** From the internal inquiry's report, quoted in "Court Action Following Erebus Disaster Inquiry," New Zealand History.

25. **"suggesting lunch, after the dust had settled"** I am indebted to retired district court judge Gary Harrison for sharing the particulars of what happened and what was said in Justice Mahon's chambers the day the friendship ended (email message to author, February 20, 2022). He was there in his capacity as junior counsel assisting the Royal Commission.

NOTES TO CHAPTER 14:
THE DEPENDABILITY OF BETRAYAL

1. **"The magnitude of the disaster," "the public importance of the issues," "the conduct of an inquiry held by a High Court judge"** These three reasons are cited in "Court Action Following Erebus Disaster Inquiry," New Zealand History.

2. **Justice Mahon informally approached** Mahon, interview, March 13, 1982, quoted in Macfarlane, EP, pp. 511–512.

3. **"he's [only] one out of five," "I'd have objected in person"** Mahon, interview, March 13, 1982, quoted in Macfarlane, EP, p. 512.

4. **"Without that, there would have been no effective opposition"** Gary Harrison, email message to author, March 9, 2022.

5. **Justice Taylor had accused the police witnesses** In an interview on March 13, 1982, with the *New Zealand Listener* (quoted in Macfarlane, EP, p. 514), Justice Mahon explained the difficulty this way:

It is said I should have warned the group of witnesses that I believed I had heard a concerted false tale about altitude [requirements] and navigation ... [that I should have given] each witness the chance to testify again to see if he [could] talk me out of that view. I would have thought that would be evidence of bias. You see, the case [hadn't] finished. This is what Judge Taylor [head of the Royal Commission into the Arthur Allan Thomas case] was doing [in his inquiry]. Taylor said [to the police witnesses], "You're a liar." So they [the police] immediately went to the High Court to try to stop the hearing on the grounds of bias.

6. **"Mahon's scrupulous attempts to be fair to the parties"** Macfarlane, EP, p. 516.

7. **a case brought in England that Mahon considered equivalent** See Macfarlane, EP, pp. 514–515, with Lord Denning and Lord Justice Lawton as quoted there.

8. **winds of change on a "fine point of law"** John Burn, "Letter to the Editor: Justice Mahon and the Erebus Disaster," *LawTalk*, New Zealand Law Society, June 12, 2020, https://www.lawsociety.org.nz/news/publications/lawtalk/lawtalk-issue-940/letter-to-the-editor-justice-mahon-and-the-erebus-disaster/. See also S. Mahon, MFS, pt. 3, "After Erebus," on "the huge wave of change occurring in administrative law" at the time of the Royal Commission hearings and the observation by a colleague of his sister, Janet, that Peter was not on the right side of it "in respect of the 'fair hearing' branch of natural justice."

9. **"was completely convinced"** Ted Thomas quoted in *Erebus Flight 901*, "Report and Reaction"; also in "Erebus Flight 901: Litany of Lies. Episode 9," *New Zealand Herald*.

Edmund Walter Thomas QC (1981), who was generally known as Ted Thomas, sat on the Law Society as a delegate for three years and served as president for one year. He then became the New Zealand Bar Association's first president (1989–1990). Afterward, he was appointed first to the High Court (1990) and then elevated to the Court of Appeal (1995) and appointed to the Privy Council (1996). Thomas retired from the Court of Appeal in 2001 but in 2005 was brought out of retirement to serve as an acting judge of the Supreme Court of New Zealand.

Like Mahon, Thomas possessed the courage of his convictions, which enabled him to adopt an unpopular course—or even go up against the establishment—where that was indicated; Phil Taylor, "New Zealander of the Year Finalist: Sir Edmund Thomas," *New Zealand Herald*, December 6, 2010, https://www.nzherald.co.nz/nz/new-zealander-of-the-year-finalist-sir-edmund-thomas/PSPXKEERIMNM4ORYCVGGSOVB3E/.

10. **"The language and tone," "if the language used is restrained"** Peter Spiller, email message to author, November 29, 2021.

11. **"creeping dislike of Woodhouse," "constant undermining of each other"** S. Mahon, MFS, pt. 3, "*Narcissus and Goldmund*," and MFS, pt. 1, "Exhibit E." In *New Zealand Court of Appeal 1958–1966: A History* (Wellington: Brookers, 2002), Spiller tellingly noted that the "Erebus case highlighted earlier tensions between Woodhouse P and Mahon J"; p. 283, n. 193.

12. **"strong personalities" and "could be blunt"** Spiller, email message to author, September 17, 2019.

13. **"the most liberal judge of," "flexible approach to legal authority"** Spiller, *Court of Appeal*, p. 110.

14. **"made up his mind long before the hearing"** Justice Mahon in a letter to Brian Todd, then the chair and managing director of Todd Petroleum Mining Co., quoted in S. Mahon, MFS, pt. 3, *"Narcissus and Goldmund."*

15. **"if Woodhouse announced there were no grounds"** Justice Mahon in papers of Margarita Mahon that she shared with Holmes; DOE, p. 388.

16. **said to possess "a seer's ability to divine"** McNeish, BR, p. 220.

17. **Justice Woodhouse was entitled to draft its judgment** Spiller, email message to author, September 17, 2019; see also Peter Spiller, "Realism Reflected in the Court of Appeal: The Value of the Oral Tradition," section 5, "Unpacking Combined Judgments," [1998] NZYbkNZJur3; (1998) 2 Yearbook of New Zealand Jurisprudence 31, http://www.nzlii.org/nz/journals/NZYbkNZJur/1998/3.html.

18. **protect "innocent people," Mahon's "obsessive" charges** Spiller, *Court of Appeal*, p. 276.

19. **Woodhouse's draft proved unpalatable, prepared to join him** Spiller, *Court of Appeal*, p. 276.

20. **"caused unfortunate tensions and strains"** Spiller, *Court of Appeal*, p. 277.

21. **wasn't thrilled but could live with** Margarita Mahon in an interview with Holmes; DOE, p. 395.

22. **"more balanced and dispassionate"** Spiller, email message to author, September 17, 2019.

23. **"bitter personal and professional blow"** Spiller, *Court of Appeal*, p. 277.

24. **"his work of a lifetime," "favorite judge of all time," emboldened "to remake the facts"** Ted Thomas in *Erebus Flight 901*, "Report and Reaction."

25. **"nothing more than a disguised approbation," merely reflected "the attitude of the management"** Justice Mahon in letter to Todd, quoted in S. Mahon, MFS, pt. 3, "*Narcissus and Goldmund.*" As late as 2012, Woodhouse would still be insisting that the chief inspector had gotten it right, the accident's ultimate cause being the pilots' decision to descend below 16,000 feet.

NOTES TO CHAPTER 15:
THE WOODHOUSE EFFECT

1. **He himself had nothing but contempt for** There are two possible reasons—not incompatible—why the Mahon Report nonetheless opens with effusions of praise for the Chippindale Report. Erebus researcher Stuart Macfarlane thinks that Mahon believed that "since he was going to call the Air NZ witnesses a bunch of liars," the public "would never believe him" if he called Chippindale a liar too (email message to author, February 11, 2022). My own view of Justice Mahon's repeated references to Chippindale's honesty and his heroic efforts to get at the truth is that he is playing a sophisticated joke of sorts on his readers. Sam Mahon famously remarked that we must find his father between the lines. I suspect that the ostensible paean to Chippindale's investigatory skills is covertly (between the lines) a repudiation of them. Why? Because the praise is fulsome—that is, excessively and insincerely extravagant.

2. **"imputations of collective bad faith"** Woodhouse, "Judgment of Woodhouse P. and McMullin J.—Delivered by Woodhouse P.," in *Judgments of the Court of Appeal of New Zealand on Proceedings to Review Aspects of the Report of the Royal Commission of Inquiry into the Mount Erebus Aircraft Disaster* (1981; Project Gutenberg, 2005; hereafter cited as WJ), https://www.gutenberg.org/files/16130/16130-h/16130-h.htm, "Judicature Amendment Act 1972," Project Gutenberg e-book.

3. **"must and do stand"** WJ, "Factual Background."

4. **"the first person who would want to know how and why"** WJ, "Intimidation of a Witness."

5. **the government's "well equipped," in-house chief inspector** WJ, "Instructions of the Chief Executive."

6. **"denied ever receiving the [briefing] material," "irrelevant"** WJ, "Specific Documents."

7. **"freak," "factor that should be taken into account," "ignorant of the deceptive dangers"** WJ, "Whiteout."

8. **had publicly expressed incredulity** On the last day of the Court of Appeal hearings, Woodhouse had inquired, "How could the crew have flown directly into the mountain when it was right there in front of them?"; in letter to Todd, quoted in S. Mahon, MFS, pt. 3, *"Narcissus and Goldmund."*

9. **"was virtually living at the McMullin residence"** Justice Mahon in letter to Todd, quoted in S. Mahon, MFS, pt. 3, *"Narcissus and Goldmund."*

10. **"He told me that he was going to resign"** Quotations from Ted Thomas in this paragraph are from *Erebus Flight 901*, "Report and Reaction."

11. **uncomfortable pressure from Prime Minister Muldoon** See Gustafson, *His Way*, p. 292.

12. **"when he became vitriolic in his criticism"** Thomas's quotations in this and the next paragraph are from *Erebus Flight 901*, "Report and Reaction."

13. **"decided me to resign," "I have made my effective protest"** P. Mahon in letter to Todd, quoted in S. Mahon, MFS, pt. 3, *"Narcissus and Goldmund."*

14. **"the minority thing"** P. Mahon in a radio interview with Sharon Crosbie, February 1, 1982, quoted in McNeish, BR, p. 255.

15. **"I don't believe in New Zealand's legal history that"** P. Mahon, interview, February 1, 1982, quoted in McNeish, BR, p. 255.

16. **"perhaps the most emotionally fraught appeal heard by the court"** Spiller, *Court of Appeal*, p. 274.

17. **"No grounds," "That was the finding that made me decide"** P. Mahon, interview, February 1, 1982, quoted in McNeish, BR, p. 255.

18. **"There is no hard and fast rule"** Spiller, email message to author, November 29, 2021.

19. **shared his "disgust at a system which would"** Sian Elias in letter to P. Mahon, quoted in S. Mahon, MFS, pt. 3, *"Narcissus and Goldmund."*

NOTES TO CHAPTER 16:
APPEAL TO JUSTICE'S FOUNTAINHEAD

1. **the institution had long been "romanticized"** Spiller, "Realism Reflected in the Court of Appeal," s. 6, "Demystifying the Privy Council."

2. **"allowed litigants in the last resort"** Spiller, *Court of Appeal*, p. 355. Often referred to as the Privy Council, the Judicial Committee of the Privy Council is, strictly speaking, actually a part of it. The panel of judges comprising it is usually five in number.

3. **"There was always a shell," "The long face," "a part of him I could never reach"** John Burn, quoted in McNeish, BR, p. 260.

4. **"inevitable" that the Privy Council would uphold, "If the government should encourage the Commissioner"** Macfarlane, EP, p. 703.

5. **"explicitly" had said so** Macfarlane, EP, p. 507. The Privy Council would later also claim the Chippindale Report made it "crystal clear"; Mahon v. Air New Zealand Ltd, s. 2.

6. **"a masterpiece of obscurity"** Macfarlane, EP, p. 508.

7. **"the operation in practice of the system of justice," "how that system deals with a major public disaster"** Macfarlane, EP, p. 702.

8. **his own approach was supported "by both justice and law"** Macfarlane, EP, pp. 702–703 (Macfarlane's phrase used in reference to Mahon).

9. **quietly called on the Law Lords in London** I am indebted to retired judge Gary Harrison, the former junior counsel assisting the Royal Commission, for this stunning intelligence (email message to author, December 14, 2021). He acquired it while serving as legal adviser for the production team of the outstanding docudrama *Erebus: The Aftermath*. Its producer, Caterina De Nave, traveled to London seeking new evidence and spoke to the clerk of the Privy Council. He informed her that Justices Woodhouse and McMullin had made separate trips to confer privately with the Law Lords.

10. **he and Attorney General McLay had quite a row** "Mahon, McLay Clash over Erebus Letter," *New Zealand Herald*, October 2, 1984, https://www.erebus.co.nz/Resources/Audio-Visual-Footage/Newspaper-Articles/ID/17/Mahon-McLay-Clash-over-Erebus-Letter.

11. **Law Lords under the chair of Lord Diplock** Not since 1948 had a case been reputed to have attracted as many attendees as this one; Holmes, DOE, p. 399.

12. **"a distant and authoritative pronouncement," never worked so hard on a judgment** Spiller, *Court of Appeal*, p. 358; p. 374, n. 35.

13. **interviewed by Christine Negroni** See Christine Negroni, *The Crash Detectives: Investigating the World's Most Mysterious Air Disasters* (New York: Penguin Books, 2016), p. 124.

14. **had a right to certain of her late spouse's paperwork** For Anne Cassin's position and feelings, see P. Mahon, VOE, pp. 218–220.

15. **"a knowingly untrue statement made by a witness"** Mahon v. Air New Zealand, s. 7.

16. **"The most disturbing element of the Erebus affair"** Macfarlane, EP, p. 703. On pp. 703–705 is a list of some twenty-seven errors and misstatements—or fudging of the facts ("economies with the truth")—made by the Privy Council.

17. **to find "as *facts* that no orchestrated litany of lies had ever existed"** Macfarlane, EP, p. 706 (emphasis added).

18. **"brilliant" analysis of the accident's causal factors** Mahon v. Air New Zealand, s. 4. The Law Lords were especially impressed by the "overwhelming case" Mahon made in his report that "the aircraft was in a 'whiteout' when it crashed" into the mountainside (s. 19, "Parenthesis on 'Whiteout'").

19. **"clears Captain Collins and first officer Cassin"** Mahon v. Air New Zealand, section 33.

20. **"was relying on the line he had himself plotted," "his own meticulous conscientiousness"** Mahon v. Air New Zealand, s. 33. See also the assessment of Paul Davison, counsel for the Collins family and NZALPA at the Royal Commission hearings, as given to Holmes in an interview (DOE, p. 393): "It was Jim Collins's conscientiousness which unseated him in the end. It wasn't a failure to respond to all the aviation aids that he had. It was his adherence to them that led him to find

himself in this insidious position where he believed he was going to be safe by adhering to his nav track and, looking ahead, of course, he was seeing what he expected to see and that's what happened."

21. **"as soloists rather than as a choir"** Justice Mahon, in Vette and Macdonald, *Impact Erebus Two* (1:02:23-1:02:45).

22. **"slipshod system of administration," "appalling blunders and deficiencies"** Mahon v. Air New Zealand, s. 34.

23. **"a series of events came together," "the company planning activity in principle"** Morrie Davis at the Royal Commission hearings, quoted in Macfarlane, EP, p. 390.

24. **"the mistake made by those airline officials"** Mahon Report, para. 393. Most accidents involve multiple factors, and in the Erebus case Justice Mahon identified ten. The crash could not have happened in the absence of any one of them, making the aircraft's collision with the mountain "a million to one chance" (para. 388). Nonetheless, the undisclosed change to the nav track was the "dominant" cause because "it was the one factor which continued to operate from the time before the aircraft left New Zealand until the time when it struck the slopes of Mt. Erebus" (para. 392).

25. **"the New Zealand Court of Appeal is," "submits that the Court of Appeal's assessment," "it would only be appropriate"** Counsel to McLay, quoted in McGee, TT, p. 252. So much for the argument that the Privy Council provided a neutral outside perspective that was necessary to prevent the

New Zealand establishment from manipulating the country's judicial processes.

26. **"Very reluctantly" they then dismissed, "the time has now come"** Mahon v. Air New Zealand, s. 34, s. 38.

NOTES TO CHAPTER 17: KNIGHTHOOD?

1. **"overshadowed by politics from start to finish"** Justice Mahon, in Vette and Macdonald, *Impact Erebus Two* (1:07:56).

2. **"raised the image of the judiciary in the public eye"** Speight, quoted in Holmes, DOE, p. 409.

3. **"The public sensed something special in him," "a very brainy man who had worked out"** Holmes, DOE, p. 396.

4. **The effort was coordinated by district court judge Anand Satyanand.** The following quotations from Ted Thomas, Carmel Friedlander, John Macdonald, and Richard Sutton are according to Holmes, DOE, pp. 408–410.

5. **"Justice Mahon was a very eminent New Zealander"** Geoffrey Palmer, in *Dominion Post*, Stuff, February 2, 2008, quoted in S. Mahon, MFS, pt. 3, "Pedagogy."

NOTES TO CHAPTER 18:
LAST YEARS

1. **Mahon had suffered a mild heart attack** S. Mahon, MFS, pt. 3, "After Erebus." See also Christopher Moore, "A Son's Sketch of Peter Mahon" (*Mainlander*, Stuff, January 31, 2009, https://www.stuff.co.nz/the-press/christchurch-life/mainlander/582343/A-sons-sketch-of-Peter-Mahon), which mentions that early symptoms of the cardiomyopathy that would kill him had appeared by the early 1980s.

2. **"difficult to get verbally close to," "a certain distance"** Maria Collins in *Erebus Flight 901*, "Report and Reaction."

3. **"I nightly pitch my moving tent"** P. Mahon, quoted in S. Mahon, MFS, pt. 5, "Epilogue." According to Pauline O'Regan, the hymn goes, "Here is the body pent / Absent from God I roam / Yet nightly pitch my moving tent / A day's march nearer home"; *Aunts and Windmills: Stories from my Past* (1991; Wellington: Bridget Williams Books, 2016), Kindle.

4. **"stunned by how emaciated Mahon was"** All quotations are from McGee, TT, chap. 14, "Dancing on the Coffins of the Dead," pp. 235–265. Anne Cassin's email of August 2, 2015, was to Erebus authority Macfarlane, who shared its contents with me (email message to author, March 11, 2022).

5. **"the devious nature of the Muldoon government's handling"** Ian Williams, "More than Rugby in this Great New Zealand Yarn," Entertainment, *Otago Daily Times*, August 16, 2008, https://www.odt.co.nz/entertainment/books/more-rugby-great-new-zealand-yarn.

NOTES TO CHAPTER 19:
FUNERAL

1. **Mahon, aged sixty-two, slipped away** Margarita Mahon's recollections in an interview with Holmes; DOE, p. 414.

2. **a stout defender of the judge** Erebus ("The Legal Process"), https://www.erebus.co.nz/Investigation/Legal-process.

3. **"run with the pack"** S. Mahon, MFS, pt. 5, "End."

4. **never became close** "We were not especially compatible spirits." The relationship "wasn't difficult, but it wasn't super-friendly either." McLay, quoted in Justin Gregory, "Crisis— Who Runs the Country?" *Eyewitness*, RNZ, June 7, 2017, https://www.rnz.co.nz/national/programmes/eyewitness/audio/201844823/crisis-who-runs-the-country.

5. **"a good man coerced"** S. Mahon, MFS, pt. 5, "End."

6. **"defended legal principle"** S. Mahon, MFS, "Appendix II."

7. **Like the earlier "man for all seasons"** An outstanding overview of this type of man is given by Robert Bolt in his preface to *A Man for All Seasons: A Play in Two Acts* (1960; Random House, 1962), pp. vii–xx.

8. **"The somber experience of mankind"** P. Mahon in a letter to the Reverend Bob Lowe (1978), quoted in S. Mahon, MFS, pt. 4, "Liberalism, Reactionism, and the Inarticulate Premise."

9. **"set like metal"** Bolt, *A Man for All Seasons*, p. xii.

10. **"refusal to back down in the face of adversity"** S. Mahon, quoted in Kimaya McIntosh, review of *Breaking Ranks*, by Sir James McNeish, *The Reader: The Booksellers New Zealand Blog*, June 1, 2017, https://booksellersnz. wordpress.com/2017/06/01/book-review-breaking-ranks-by-sir-james-mcneish/.

11. **"learned as a child never, ever to"** S. Mahon, quoted in Jarod Booker, "Son Crafts a Portrait of Courage," *New Zealand Herald*, August 22, 2008, https://www. nzherald.co.nz/nz/son-crafts-a-portrait-of-courage/ YA6EZFUCJJXRIWX3JZ5TZ7LY6Q/.

12. **"He was a tough man"** S. Mahon, quoted in Bruce Ansley, "Out of the Shadows," *New Zealand Listener*, September 6, 2008, pp. 24–26.

13. **"drew very, very severe lines"** S. Mahon, quoted in Booker, "Son Crafts a Portrait of Courage."

14. **"It suited his wit"** S. Mahon, MFS, pt. 2, "Album," "Murder."

15. **"he quietly played the pieces"** S. Mahon, quoted in Booker, "Son Crafts a Portrait of Courage."

16. **"the *making* of him"** S. Mahon, quoted in Ansley, "Out of the Shadows."

17. **"He was an artist in his way"** S. Mahon, MFS, pt. 1, "Lago Como, Italy."

18. **"by which justice can be achieved," "example of the law working"** Paul Davis, quoted in Clare de Lore, "I Have Always Had a Sense of Justice," *New Zealand Listener*, October 8, 2015.

NOTES TO CHAPTER 20:
MORE LIVES LOST, 1980–1989

1. **the accident at Canada's Dryden Municipal Airport** My account of the Commission of Inquiry is from Commissioner Virgil P. Moshansky, foreword to *Beyond Aviation Human Factors: Safety in High Technology Systems*, by Daniel E. Maurino, James Reason, Neil Johnston, Robert B. Lee (1995; United Kingdom: Taylor & Francis, 2016).

2. **Overnight, the Mahon Report changed the prospects** Jim McLay in *White Silence*, ep. 4, "Love Song."

3. **tract was essentially a paean to** Charles Henry Noel L'Estrange, *The Erebus Enquiry: A Tragic Miscarriage of Justice* (Auckland: Air Safety League of New Zealand, 1995).

4. **"deeply involved in what went on"** Maurice Williamson, quoted in David Farrar, "Maurice's Valedictory," *Kiwiblog*, August 11, 2017, https://www.kiwiblog.co.nz/2017/08/maurices_valedictory.html.

5. **"dancing around the Maypole," "somebody the caliber of Mahon"** Williamson, in *Erebus Flight 901*, "Report and Reaction."

6. **"Justice Mahon got it right," "so wrong to blame just the pilot," "systemic failure, not one error"** Williamson, quoted in Farrar, "Maurice's Valedictory."

7. **"one of New Zealand's true heroes"** Williamson, quoted in "Erebus Report Finally Recognized," New Zealand Government Press Release, Scoop, August 25, 1999, https://www.scoop.co.nz/stories/PA9908/S00427.htm.

8. **attempted to get the Court of Appeal's two judgments along with** "Ongoing Debate Over Erebus," New Zealand History, updated September 18, 2019, https://nzhistory.govt.nz/media/photo/neverending-debate-over-erebus.

9. **did not appreciate the day of the ceremony** When Williamson belatedly learned of the issue, he was no longer in a position to do anything to rectify the situation; Andrew Laxon, "Minister Backs Clearing Erebus Pilots' Names," *New Zealand Herald*, September 9, 2011, https://www.nzherald.co.nz/nz/minister-backs-clearing-erebus-pilots-names/FVDQFQYH6PWSTYUZR4B24YSYKE/.

10. **the Mahon Report is not to be found within Annex 13** Kurt Bayer, "Government Reject Erebus Disaster Report Claims," *New Zealand Herald*, November 30, 2012, https://www.nzherald.co.nz/nz/government-reject-erebus-disaster-report-claims/63FINK2RSI2562CE6L7PTKRYCA/.

11. **"probably ten years ahead of its time," "certainly the Dryden Report"** See International Civil Aviation Organization, *Human Factors Digest No. 10: Human Factors, Management and Organization* (Montreal: ICAO, 1993), available for purchase from ICAO.

NOTES TO CHAPTER 21:
THE NATURE OF COURAGE

1. **"it seems like history is repeating itself"** Peter Mahon's sister, quoted in *White Silence*, ep. 6, "White Silence."

2. **"My father was the bravest person"** S. Mahon, quoted in Booker, "Son Crafts a Portrait of Courage."

3. **"never be pushed, never give way, never compromise"** S. Mahon, quoted in Ansley, "Out of the Shadows."

4. **"They picked the wrong man to try to squeeze"** S. Mahon, quoted in Booker, "Son Crafts a Portrait of Courage."

5. **"I see it as a great blot"** M. Mahon quoted in Jane Phare, "Erebus Milestone Stirs Up Emotions," *New Zealand Herald*, October 17, 2009, https://www.nzherald.co.nz/nz/erebus-milestone-stirs-up-emotions/LY33MS4NMQNPUVET XKMPFYT22E/.

6. **"I really had no alternative"** P. Mahon, quoted in McNeish, BR, p. 243.

7. **executed his task as Royal Commissioner as ably as, What greatly disillusioned him** Phare, "Erebus Milestone Stirs Up Emotions."

8. **It's reported** Greg McGee, TT, pp. 241-242; this includes the relevant passage from *The Divine Comedy* ("Inferno," canto 31).

9. **"seven years to run as their top pilot"** Vette, quoted in "Court Action Following Erebus Disaster Inquiry," New Zealand History, and *Impact Erebus Two*.

10. **earning him praise** The praise quoted here was that of Professor R. A. McD. Galbraith in the short speech he made on July 16, 1988, requesting the vice-chancellor to bestow the degree of honorary Doctorate in Engineering on the intellectually gifted Captain Vette; see Vette and Macdonald, *Impact Erebus Two* (1:12:57-1:13:35), and Chris, "Capt. Vette (Ex-Chief Pilot of Air New Zealand) Honored," Society Culture New Zealand, Google Groups, August 6, 1998, https://groups.google.com/g/soc.culture.new-zealand/c/-2OOr0eOSj4.

11. **Continuing his research on crashes into terrain** Juliet Rowan and Mathew Dearnaley, "Queen's Birthday Honours: Captain Gordon Vette, J. P.," *New Zealand Herald*, June 4, 2007, https://www.nzherald.co.nz/nz/iqueens-birthday-honoursi-captain-gordon-vette-jp/D3NGRFR3YY2XY4OLXYZZPDRT6Y/. See also, Capt. G. A. Fernando, "And They Called it Pilot Error," Kathryn's Report, December 21, 2014, http://www.kathrynsreport.com/2014/12/capt-g-fernando-and-they-called-it.html.

12. **By 2007 terrain awareness and warning systems (TAWS)** Wikipedia, s.v. "Gordon Vette," "Legacy," last updated February 3, 2022, https://en.wikipedia.org/wiki/Gordon_Vette#Legacy

13. **"a final congratulations from the country"** Mark Vette, quoted in Rowan and Dearnaley, "Queen's Birthday Honours."

14. **New Zealand Special Services Medal (Erebus) would be awarded** Prime Minister Clark's comments are quoted in a press release titled "Erebus Special Services Medal to Be Awarded," November 29, 2006, https://www.beehive.govt.nz/release/erebus-special-service-medal-be-awarded

15. **known as *Erebus: Operation Overdue* as well as *Erebus: Into the Unknown*** Purdy and Burger.

16. **"a hellish story told hellishly well"** Colin Hogg of the *New Zealand Herald*, quoted in "Erebus Operation Overdue," Rogue Productions, https://rogueproductions.co.nz/productions/details/165/Erebus-Operation-Overdue.

17. **"Gray-haired vision-impaired lady with Parkinson's disease," "especially hands-on sailing"** Quoted in "About Sailing," Sailibility Nelson, https://www.sailabilitynelson.org.nz/about-sailing

NOTES TO CHAPTER 22:
THE ANSWER TO LORD DIPLOCK'S QUESTION

1. **"a public disaster carrying highly embarrassing overtones"** Macfarlane, EP, p. 702.

2. **"the accident aircraft was 'left of track'"** Andrew Laxon, "Air NZ'S Missing Flight Path Evidence Confirms 'Orchestrated Litany of Lies' over Erebus – Judge Gary Harrison," *New Zealand Herald*, November 27, 2019, https://www.nzherald. co.nz/nz/air-nzs-missing-flight-path-evidence-confirms-orchestrated-litany-of-lies-over-erebus-judge-gary-harrison/ FKIUCGZ6LTZTNTBQ5WGDHAZBGY/. The crash site was below the caldera of a subsidiary cone of Mt. Erebus.

3. **the airline later rewarded Greenwood** Holmes, DOE, p. 302. "Managerial pressure" was explained this way to Justice Mahon by a seasoned company attorney exasperated that the Court of Appeal "could be so naïve" regarding the assessment of evidence given by a series of witnesses from a *single* organization:

> Can they not understand the pressures that are always on company witnesses where their evidence is briefed possibly by an internal company lawyer, where those briefs are no doubt reviewed by senior executives in the organization, where inevitably there are pressures as to loyalty, not rocking the boat, thinking of the future of

the organization [and an expectation by the senior executives of the company that individuals will go along with the company line for the sake of what those executives] see as the interests of the company and with the implicit understanding created in the mind of the employees that their prospects with the company would be affected if they did not go along with the expected approach?

Justice Mahon quotes the seasoned company lawyer in a letter to Justice Laurie Greig on February 2, 1982, that is reproduced in Macfarlane, Unpublished Papers ("Erebus in Court," p. 18).

4. **part of the conspirators' plan for influencing** The offices of Anne Cassin's counsel, Roger Maclaren, were in the same building as those of well-connected attorney Lloyd Brown. He believed that Brown's selection to represent Air New Zealand was also part of the conspirators' plan. Rumors had begun circulating in the building that the power lawyer's Midas touch of turning cases into lucrative legal exercises was slipping away. (He had recently lost several cases through no fault of his own.) If the rumors had any substance, that could have put extra pressure on him to succeed in his next case. Maclaren himself suspected that the airline felt that Brown would be so highly motivated that he'd be willing to do whatever was necessary to succeed, including using tactics he would not normally use. Justice Mahon may not have been alive to this possibility initially, but partway through the Royal Commission hearings he would have realized that his old friend was masterminding strategy for lying airline witnesses; Macfarlane, Unpublished Papers ("Erebus Events," pp. 15–16).

5. **minutes of an Air New Zealand board of directors' meeting** P. Mahon, VOE, p. 272.

6. **"Mahon's logic was impeccable"** McLay, quoted in *Erebus Flight 901*, "Report and Reaction."

NOTES TO CHAPTER 23:
APOLOGY

1. **"In a century or two, there will be an apology"** Quoted remarks by Maria Collins, Prime Minister Jacinda Ardern, and Margarita Mahon are according to *White Silence*, ep. 8, "The Apology." For the full text of Prime Minister Ardern's speech, see "Prime Minister Delivers Erebus Apology," Latest News, Erebus, November 29, 2019, https://www.erebus.co.nz/Home/News-Page/ID/543/Prime-Minister-Delivers-Erebus-Apology.

2. **"only three of us left now" to explain the accident correctly** Peter Grundy, in "Erebus Flight 901: Litany of Lies. Episode 6: Former Top Air NZ Pilot Insists – 'It Was Pilot Error,'" *New Zealand Herald*, November 21, 2019, https://www.nzherald.co.nz/nz/erebus-flight-901-litany-of-lies-episode-6-former-top-air-nz-pilot-insists-it-was-pilot-error/DMWELET3QZGXC7SXTYN3JR7GSQ/. Captain Grundy was intimately involved in the planning of the airline's Antarctic aerial tours.

SELECTED BIBLIOGRAPHY

How to access documents cited in the notes is indicated
on the first occasion of each one's use. Here are thus
listed only salient official documents, certain audiovisual
materials, and books I found particularly pertinent
to the subject matter of *Judgment on Erebus*.

ESSENTIAL RESOURCES &
OFFICIAL REPORTS

Erebus: The Loss of TE901
New Zealand Air Line Pilots' Association. Erebus. N.d. https://
www.erebus.co.nz/.

The official Erebus website hosted and maintained by the New
Zealand Air Line Pilots' Association (NZALPA), which calls
it "the most comprehensive source of information on the 1979
Erebus disaster." A stunning collection of Erebus materials of
various types—all in one place.

Chippindale Report
*Aircraft Accident: Air New Zealand McDonnel-Douglas DC10-30 ZK-
NZP, Ross Island, Antarctica, 28 November 1979, Report 79-139.*

Office of Air Accidents Investigation, Ministry of Transport, 1980. Digital copy. https://www.erebus.co.nz/Portals/4/Documents/Reports/Chippindale/79-139%20Chippindale%20Report%20-%20Web.pdf.

Mahon Report

Report of the Royal Commission to Inquire into the Crash on Mount Erebus, Antarctica, of a DC10 Aircraft operated by Air New Zealand Limited. P.D. Hasselberg, Government Printer, 1981. Digital copy. https://www.erebus.co.nz/Portals/4/Documents/Reports/Mahon/Mahon%20Report_web.pdf?ver=2019-06-17-124911-650.

Court of Appeal Judgments

Judgments of the Court of Appeal of New Zealand on Proceedings to Review Aspects of the Report of the Royal Commission of Inquiry into the Mount Erebus Aircraft Disaster. Project Gutenberg, 2005. https://www.gutenberg.org/files/16130/16130-h/16130-h.htm. Project Gutenberg e-book.

This text contains the majority judgment by Cooke, Richardson, and Somers JJ and the minority judgment of Woodhouse P and McMullin J.

Privy Council Judgment (*Mahon v. Air New Zealand*)

Judgment of the Lords of the Judicial Committee of the Privy Council, Delivered on the 20th October 1983. Mahon v. Air New Zealand Ltd. [1984] 3 All ER 201; [1983] NZLR 662. http://www.nzlii.org/nz/cases/NZPC/1983/3.html.

Cockpit Voice Recorder Transcripts
NZALPA. "The CVR Transcript." Erebus. N.d. https://www.erebus.co.nz/The-Accident/Transcript.

The National Transportation Safety Board's and Chief Inspector Chippindale's transcripts can be found here.

AUDIO & VIDEO

Videos

Erebus: The Aftermath
Sharp, Peter, director; Greg McGee, writer; and Caterina De Nave, producer. *Erebus: The Aftermath*. New Zealand: Television New Zealand (TVNZ), 1987.

Uploaded to YouTube as "'Erebus—The Aftermath' Part One," https://www.youtube.com/watch?v=VImFx0GrjHE&t=1s, and "'Erebus—The Aftermath' Part Two," https://www.youtube.com/watch?v=avnBAtLvVqY&t=5s. John Parker on June 21, 2016.

Riveting recreation of the Erebus story.

Erebus 901 – The Secret Files
Graham, Jill, director and producer. *Secret New Zealand*. "Erebus Unsolved." New Zealand: Greenstone Pictures, 2003.

Uploaded to YouTube as "Erebus 901—The Secret Files." John Parker on August 27, 2019. https://www.youtube.com/watch?v=DpOeL9qkKmY.

Forensic documentary that explains how Captain Collins's empty ring binder came to represent all that was enigmatic or indeed sinister about the investigation into what caused TE901's crash.

Impact Erebus Two

Vette, Gordon, and John Macdonald. *Impact Erebus Two*. Auckland: Aviation Consultants, 1999. Introductory DVD.

Uploaded to YouTube as "Impact Erebus Two" (single video). mayorofthenonsense on June 15, 2013. https://www.youtube.com/watch?v=xyWvOI_MD-Q.

Uploaded to Erebus, "Audio & Video Gallery," as "Impact Erebus Two" (nine video clips). https://www.erebus.co.nz/Resources/Audio-Visual-Footage/Audio-Video-Gallery.

Fascinating, multifaceted video that includes in-depth interviews by John Macdonald with Gordon Vette as well as an arresting interview of Vette and Mahon together, visual demonstrations of Vette's discoveries with respect to mental sets and the mechanics of human perception, and other important material.

Lookout: The Mt. Erebus Disaster

Aberdein, Keith, writer, and John Keir, producer and director. *Lookout.* "Flight 901 – The Erebus Disaster." New Zealand: TVNZ, 1981. TV series.

Uploaded to YouTube as "TVNZ's 'Lookout': The Mt Erebus Disaster." mayorofthenonsense on June 15, 2013. https://www.youtube.com/watch?v=yP36X0BsMQ0.

The first in-depth documentary on Erebus, directed and produced by John Keir. Keir's work includes valuable footage from the Royal Commission of Inquiry hearings.

Podcasts

Erebus Flight 901: Litany of Lies?

McAlpine, Gary, and John Keir. *Erebus Flight 901: Litany of Lies?* November 2019. Produced by New Zealand Media and Entertainment. Podcast. MP3 Audio. Archived by the National Library of New Zealand. https://natlib-primo.hosted.exlibris-group.com/primo-explore/fulldisplay?docid=NLNZ_AL-MA11350355710002836&context=L&vid=NLNZ&search_scope=NLNZ&tab=catalogue&lang=en_US.

White Silence

Wright, Michael, and Katy Gosset. *White Silence.* November 2019. Produced by Stuff and Radio New Zealand (RNZ). Podcast. https://shorthand.radionz.co.nz/white-silence/.

Both podcasts are extremely well done and totally engrossing. They suffer from a serious interpretative error, however, of which listeners should be aware. Both split the blame for the Erebus disaster between the dead pilots and Air New Zealand when Mahon and Vette had, several decades earlier, conclusively proved that the cause of and culpability for the air accident lay totally with the airline.

Audio Recordings

NZ Biography: Peter Mahon

Morris, Grant. "NZ Biography: Peter Mahon." Interview with Jesse Mulligan. *Afternoons with Jesse Mulligan*. RNZ. December 19, 2019. https://www.rnz.co.nz/audio/player?audio_id=2018727693/nz-biography-peter-mahon.

In an overview of Peter Mahon, Victoria University of Wellington's Dr. Grant Morris expresses his initial astonishment at learning that while outside the legal profession Mahon was considered a hero, for the legalistic minds inside it he was deemed a judge who erred, went too far, and paid the price. Morris concludes the talk by observing that Mahon's speaking truth to power (especially since he was part of the establishment himself) makes Mahon's story "one of the most interesting and inspiring" in New Zealand's legal history.

BOOKS

Allinson, Robert Elliott. *Saving Human Lives: Lessons in Management Ethics.* Netherlands: Springer, 2005. Allinson devotes a meaty chapter ("The Disaster on Mt. Erebus," pp. 256–321) to the aftermath of TE901's demise. His purpose is to compare the Chippindale Report to the Mahon Report to determine which is fraudulent and which is correct.

Hickson, Ken. *Flight 901 to Erebus.* Christchurch: Whitcoulls Publishers, 1980. Hickson was working for Air New Zealand when he wrote and published this book, which is full of praise for Chief Inspector Ron Chippindale and extols the correctness of his account of the cause of and culpability for the Erebus disaster. Putting his political objectives to one side, Hickson also shares interesting, pertinent background information that this author has not encountered elsewhere.

Holmes, Paul. *Daughters of Erebus.* Auckland: Hodder Moa, 2011. The late New Zealand broadcaster Paul Holmes broke new ground by writing the first book-length narrative account of the Erebus accident and its highly contentious aftermath. One valuable feature of his study is that he interviewed individuals still alive at the time of the thirtieth anniversary of the disaster but deceased by the fortieth. Another is that Holmes also received important assistance from top Erebus authority Stuart Macfarlane, who improves all texts he touches. Stylistically, however, the work is repetitive and disjointed, with distracting digressions. It is also confusing because of the author's ill-advised leaps forward and backward in time. Perhaps reflecting Holmes's own controversial manner, the book's tone is strident. Despite these weaknesses, *Daughters of Erebus* is in the end a valuable resource.

Macfarlane, Stuart. *The Erebus Papers: Edited Extracts from the Erebus Proceedings with Commentary.* Auckland: Avon Press, 1991. This massive tome is a must-have resource for Erebus researchers because it both reproduces all the most vital extracts relating to the Erebus proceedings and rigorously critiques them.

Mahon, Peter. *Verdict on Erebus.* Auckland: William Collins Publishers, 1984. Justice Mahon reprises major themes of his official report in this highly readable, award-winning popular account of his experience as head of the Royal Commission of Inquiry into the Erebus disaster.

Mahon, Sam. *My Father's Shadow: A Portrait of Justice Peter Mahon.* Longacre Press, 2008; reprinted 2009. Kindle. Artist Sam Mahon paints an impressionistic portrait of his famous father in a series of vignettes that illuminate key aspects of the judge's temperament and character.

McGee, Greg. *Tall Tales (Some True): Memoirs of an Unlikely Writer.* New Zealand: Penguin Books, 2008. While writing the script for the mesmerizing docudrama *Erebus: The Aftermath* (1987), McGee became well acquainted with Justice Mahon, whom he interviewed repeatedly in 1986 as part of his research for the project. In *Tall Tales* (chapter 14, "Dancing on the Coffins of the Dead," pp. 235–266), McGee describes what it was like working on a politically charged project and, in the process, getting to know the ailing judge during the last full year of his life.

McNeish, James. *Breaking Ranks: Three Interrupted Lives.* New Zealand: HarperCollins, 2017. The critically acclaimed author devotes the last third of his study to an insightful assessment of Justice Mahon's character. I thank Sir David Baragwanath, erstwhile senior counsel assisting the Royal Commission, for bringing this work to my attention as I commenced my Erebus research.

Myles, Sarah. *Towards the Mountain: A Story of Grief and Hope Forty Years on from Erebus.* New Zealand: Allen & Unwin, 2019. Kindle. This book by the granddaughter of one of TE901's crash victims contains (chapter 8) an excellent synopsis of the protocols and procedures in place during Operation Overdue in both its phases: first at the Antarctic accident site and later at Auckland University's morgue.

Negroni, Christine. *The Crash Detectives: Investigating the World's Most Mysterious Air Disasters.* New York: Penguin Books, 2016. A veteran aviation journalist devotes a chapter (part 2, pp. 117-131) of her splendid book on notably enigmatic air disasters to an excellent critical summary of the Erebus story. Negroni seems to have been the last (along with Stuart Macfarlane) to have corresponded with first officer Cassin's widow (2015). Again Anne expressed her outrage that vital flight documents were taken from her home without her consent or even knowledge by an Air New Zealand pilot sent to her home.

Palin, Michael. *Erebus: One Ship, Two Epic Voyages, and the Greatest Naval Mystery of All Time.* Vancouver: Greystone Books, 2018. In the first half of this wonderfully informative and well-written work, the versatile former Monty Python star describes Captain James Clark Ross's historic discovery of the slice of Antarctica in which Air New Zealand later operated its sightseeing tours.

Vette, Gordon, and John Macdonald. *Impact Erebus.* New York: Sheridan House, 1984 (originally published in New Zealand in 1983). Going up against his colleagues and boss at Air New Zealand, Captain Vette worked tirelessly to ascertain how a pilot of Captain Jim Collins's caliber could crash his jetliner into an Antarctic volcano while flying in broad daylight and clear air. His exhaustive research into mental sets and the mechanics of human visual perception, both lucidly explained in meticulous detail in *Impact Erebus,* enabled him to establish the accident's proximate cause. Attached to the text are valuable appendices by various authors with specialized knowledge bearing on the accident.

INDEX

Page numbers in *italic* indicate maps; those in **bold** indicate photos; those containing n indicate notes.